To Armageddon and Beyond

Ransom R. Radford

P.O. Box 47
4384 Moultrie Rd.
Camilla, GA 31730

To Armageddon and Beyond
Ransom R. Radford

ISBN 0-9716029-1-3

Quill Publications
P.O. Box 8193
Columbus, Georgia 31908

Dedication

I dedicate this work to the gentle memory of my Mother, Nora Bass Radford, who loved to have her children read to her in the evening. At age ten, I read the entire Book of Revelation to her at one setting.

Ransom R. Radford

Table of Contents

Preface

The purpose of this Book is to help those who read it understand the times in which they are living. Everyone can sense that things are changing, indeed. Things will never be the same again, especially since 9-11. Change is taking place in many arenas. Political strategies are changing world-wide. So are the Social Orders and cultures of the world. The top three areas of change could be listed as the Political, the Social and the Religious. The International Community does not feel as secure as it once did. There is not as much trust. The demands of modern technology are pressing in upon us. The fast lane of society is much more prevalent. Wars are fought differently. Space exploration is expanding. Religious thought and theologies are not the same as they used to be. All of these things, working together, are changing the mind-set of the International Community.

It is in times like these that Jesus will return and receive His Church unto Himself. It is also in times like these a World Leader will emerge, seemingly with all the answers The world will call him a genius, but the Bible calls him the antichrist. His days are already numbered. His rise and fall is already set. The Bible is very specific on this.In the days before his appearing, nothing will seem to have any certainty. There will be no absolutes.

Everything will be relative.

But there is a new world coming, a new Kingdom, and Christ will be its King. The watchword for this coming Kingdom is HOLINESS UNTO THE LORD. There will be no sin and its people will return to the perfection that Adam had before The Fall.

It is the author's prayer that the readers of this book will find their special place in the Lord's Kingdom and live in communion and fellowship with Him every day.

Chapter One

The Present Church Age

And do not realize and understand that you are wretched, pitiable, poor, blind and naked. (Rev. 3:17 Amplified Bible)

After observing the world-system in the beginning of the 21st century, any sincere student of the Bible will have to agree that this generation of humanity is living in the last days. This writing is about what is happening right now in the world in which we live and also, things that are about to happen.

I mentioned above the world system. Many centuries ago, a generation or two after the flood, Nimrod and his generation began building the world system. It's recorded in the 11th chapter of Genesis. They said, "Let us build a city with a tower that reaches unto the heaven." That tower today indeed reaches into the heavens and will play a vital part in the events of the Great Tribulation Period. It is called the Internet, the World Wide Web. While the Internet is not evil within itself, it is abused by those who use it for unethical and immoral purposes. The same is true with any other invention or technology that man may bring about. These things can be used for good or evil.

If we search the Scriptures carefully, we will find that Satan always attacks mankind through his religious beliefs. To do this, Satan must get into the mind of the individual and convince him that his suggestions and reasonings are right. He used the Scriptures in his temptations of the Lord. "If you be the Son of God," he said (trying to interpose doubt), "make these stones turn into bread and eat." You see, this was after Jesus had fasted forty days and nights, and He was hungry. Satan was saying to Jesus, "Use your God-given power and work a miracle." Satan knew that Jesus was the Son of God and Jesus knew He was the Son of God. Satan's goal was to entice Jesus to use the power of God to satisfy His physical need. When Jesus refused, Satan took Him to the pinnacle of the Temple and said, "If you be the Son of God, (again interposing doubt) cast yourself down, for it is written, He shall bear thee up lest thou dash Thy foot against a stone." It was the time of the morning sacrifice and there was a great number of people in the Temple Court praying for the coming of the Messiah. Satan was abusing the Scriptures and mis-applying them. He was misconstruing God's Word. The temptation is, "While the Jews are praying for you to come, jump down and float into their midst. They will immediately proclaim you the Messiah." This would have been a great miscarriage of the Truth of God. If Jesus had yielded to this temptation, the cross would have been unnecessary. This was Satan's goal: to get Him to bypass the cross, and thereby forfeit the redemption of man.

Of course the temptation was three-fold. Satan was

allowed to take Jesus to an exceedingly high mountain and in a moment make all the kingdoms of the world pass before Him. "All these things," Satan said, "I will give you if you will fall down and worship me." This is Satan's goal, to get mankind to worship him. There is nothing that he will not try in his efforts to achieve that goal.

In these last days, Satan is more and more closing in on man's belief system. He is presenting prosperity as a sign of God's favor. He has convinced the Church-world to 'use their faith' in order to get God to give them what they want and consequently be saying, "I am rich and increased in goods and have need of nothing." That is a self-description of the worldly church. Jesus' description of the Last-Day-Church-World is quite different. He says it is wretched, and miserable, and poor, and blind, and naked and it doesn't even know it. While the shouts and dancing and swoonings and tongues and interpretations are going on, there is a lack of sound doctrine. But Jesus has the answer for this Laodicean Age.

The word "Laodicea" itself means "ruled by the people" or "democratic." Jesus' instruction to the Church is to repent. That means "change your mind." This Church Age must realize the circumstances it is in and come back to the altar of repentance and be "truly rich." It must experience the life-changing, life-giving power of the Holy Ghost that comes with true conversion. When this happens they will not only have "changed minds," but also changed spirits and changed goals and motives. This Church Age must buy white raiment from Him; "white raiment, that the shame of your nakedness is not manifested." That is, that they be not caught trying to serve God in their natural sin nature and carnality. They must be sanctified. They must set themselves apart for service to God. Once this happens, God will sanctify them in a soul-cleansing work of grace. Regeneration is a divine work of grace. So is sanctification. In the Bible the Greek word for "holy""or "holiness""is "hagios." It is translated many times as "sanctify" or "sanctification." Now we know that holiness belongs to God. In fact, it is one of God's chief attributes. Let me say here that man has no holiness of his own. And holiness is not demonstrated by the cut of one's clothes or the ornamentaion, or lack of it, with which one may adorm himself or herself. Holiness is a principle of divine life belonging to God only. But God is willing and able to impart a measure of holiness or sanctification to every true believer. Jesus prayed, " Sanctify them through Thy truth, Thy word is truth."John 17:17

Words cannot adequately describe the power of the new birth, or sanctification, or the baptism of the Holy Ghost, but these are the works of grace that Jesus urges this Laodicean Church Age to obtain from Him. Allow me to say something about the Baptism with the Holy Ghost. It is not someone teaching another to speak in a few unintelligible syllables. It cannot be described as "speaking in tongues" alone. Christ is the Baptizer. He does the baptizing. When the Holy Spirit comes in, He speaks for Himself. It is a precious experience for the believer. You see, the third Person of the Godhead comes into the soul, spirit, and mind of the believer and takes up His residence within that believer. It is a life-changing experience within itself. This is what Jesus is saying in the case of the Laodicean Church. He is urging the

Church to buy of him these three things: the experience of regeneration, sanctification, and the indwelling presence of the Holy Spirit.

A mistake this Last Day Church Age is making is teaching that Joel's prophecy concerning the outpouring of the Holy Spirit is a thing of the future, even of this Church Age. But it has already happened. It happened on the Day of Pentecost and is recorded in detail in Acts Chapter 2. Peter said so! When the skeptics made their charge, "These men are drunk with new wine," Peter stood up in the Spirit and refuted that notion. "These men are not drunk, as ye suppose, but this is that spoken by Joel, the prophet, that in the last days God will pour out of His spirit on all flesh..." Let's consider for a moment how God gave the Law. You can read about it in Exodus, Chapter 19. "And Mount Sinai was altogether on a smoke, because the Lord descended upon it in fire: and the smoke thereof ascended as the smoke of a furnace, and the whole mount quaked greatly." V. 18. In Exodus Chapter 20, God gave the Ten Commandments. But he never came down on Mount Sinai and spoke to Israel again in that fashion. Likewise, He is not going to initiate another Day of Pentecost.

The Holy Ghost has already been given and Jesus gives the remedy for this Last Day Church Age as He addresses the Laodicean Church. The Church through its misunderstanding of what God has done and is doing is boasting about its wealth in worldly goods, and thinks it is spiritually rich. In so doing it has turned the Lord away and we see Him standing outside the Church, knocking on a door. But it is not the Church door. It is the door of the individual heart. His plea is not to the

Church, but to the individual. He is saying, "If any man hear My voice and open the door, I will come into him and sup with him and he with Me."

There is a vast difference between where the Church is and where it could be. We can determine this by history. History records, "But the miraculous cure of diseases of the most inveterate or even preternatural kind can no longer occasion any surprise, when we recollect that in the days of Irenaeous, about the end of the second century, the resurrection of the dead was very far from being esteemed an uncommon event; that the miracle was frequently performed on necessary occasions, by great fasting and joint supplication of the church of the place, and that the persons thus restored to their prayers had lived afterwards amongst them many years." (Edward Gibbons, *The Decline and Fall of the Roman Empire.* Page 161.)

This is very meaningful information, especially since Gibbons was a secular historian. His intent was not necessarily to lift up the Church or commend its belifs and practices, but to present history as it actually happened. This is the chief aspiration of all true historians. This description of the early Church is a far cry from the Church described by the Lord as the Laodicean Age.

The Church did not retain the white-heat spirit of revival for long. It did for a little over 300 years. This was during the Ephesus "First Love" Age and the Smyrna "Suffering" Church Age. Great persecution of the Church was going on during this time. But, then, the Emperor Constantine was converted and things began to change. Rome ceased its persecution of the Christians.

Their confiscated land and possessions were restored and the character of the Church changed. Let history speak again: "Since any friend to revelation is persuaded of the reality and every reasonable man is convinced of the cessation of miraculous powers, it is evident that there must have been some period in which they were either suddenly or gradually withdrawn from the Christian church. Whatever era is chosen for that purpose, the death of the Apostles, the conversion of the Roman Empire, or the extinction of the Arian heresy, the insensibility of Christians who lived at that time will equally afford a just matter of surprise. They still supported their pretensions after they had lost their power. Credulity performed the office of faith; fanaticism was permitted to assume the language of inspiration, and the effects of accident or contrivance were ascribed to supernatural causes." (Edward Gibbons, *The Decline and Fall of the Roman Empire*. Page 162.)

God is calling this Church of the Last Days to repentance, and sanctification, and the genuine infilling of the Holy Spirit. All three are divine works of grace. There are a lot of counterfeits out there. We must remember there cannot be a counterfeit unless there is first the real. It is Christ Who stands outside the door and knocks. It is the door of the individual heart that is portrayed here, not the door of the Church. The door can only be opened from the inside and that by the individual himself. John said, "I entered into a different kind of experience in the sphere of the Spirit [His absolute control] on the Lord's day." Rev. 1:10. (The Expanded Translation.) Counterfeits do not work. They only deceive. Dear Reader, you can have the real. Don't let anyone deceive you! The truly saved needs no one to convince him that he is saved. He knows it! The sanctified man needs no one to convince him that he is sanctified. He knows it! The Spirit filled man needs no one to convince him that he is filled with the Spirit. He, also, knows it!

Chapter Two

The Coming of the Lord

I will come again and receive you unto Myself; that where I am, there ye may be also. (John 14:3b. KJV)

There are two stages to the Second Coming of Jesus. The first is the Rapture. No one will visibly see Him at that time. Jesus told His disciples that He woul come as the lightning shines from the east to the west. Matt. 24:27. Paul spoke of the Rapture in this manner, "Behold, I show you a mystery; we shall not all sleep, but we shall be changed, in a moment, in the twinkling of an eye, at the last trump: for the trumpet shall sound, and the dead shall be raised incorruptible, and we shall be changed." I Cor. 15:51, 52. Jesus is going to take His Church, His Bride out of this troubled world so that she can forever be with Him. He loves His Church, His Bride, and she loves Him. The second stage of the Second Coming of Jesus is visible. The place is Armageddon. The event is recorded in Rev. 19:11-16. Many International world events must take place between the first stage and the second stage of His Second coming.

When Jesus catches His Bride, His holy Church, away, it will be a day like today has been, or a night like last night was, just a normal day or a normal night, but it will be an absolute spiritual event. Those who have believed in Him as the Scriptures have taught will be changed in a moment, in a flash, in the twinkling of an eye. The believers who have died will be first. Their corruptible bodies will instantly be changed to a new body, like unto His own resurrected body, and rise to be united with the soul and will forever remain with Him. Then the living believers will undergo the instant change and join them in the air.

The Rapture will occur at the end of the Church Age which has existed from the time of His Ascension and lasts until the Rapture. John, in the Book of Revelation, becomes a type of the Bride in Rev. 4:1. He says, "After this I looked, and behold, a door was opened in heaven; and the first voice which I heard was as it were a trumpet talking with me; which said, Come up hither, and I will show thee things which must be hereafter. And immediately I was in the spirit: and, behold, a throne was set (or being set) in heaven, and one sat on the throne." Rev. 4:1, 2. The remainders of Chapters 4 and 5 describes heavenly scenes. How the raptured saints worship and praise Him! In their Rapturous intellect they understand more fully than ever before their salvation. The raptured company is great. In fact, it is innumerable and consists of those from "every kindred, and tongue, and people, and nation." Rev. 5:9. How happy they are! And how they praise Him! How they shout and sing and honor Him! This is the heavenly scene immediately following the Rapture. The Rapture is ever-imminent. No other Scripture has to be fulfilled. It can happen at any moment.

The 6th Chapter of Revelation begins with the opening of the seven seals. The scene is transferred from heaven to the earth. The white horse and rider of the first seal symbolizes an apostate revival of religion on the earth. The breaking of the second seal brings a red horse and rider. We want to look closely at these two horses and their riders now.

First, the white horse and its rider. It is true that the horse is white. Some Bible scholars take this to symbolize a holiness revival. The rider they take to be Jesus Christ. The color of the horse does symbolize a revival of sorts, but it is an apostate revival. It is a time of intermingling of false doctrines with the truth. It is a mixture of religious beliefs and teachings, some of them Scriptural and some of them not. The description of the rider does not fit that of our Lord. It is stated that "he had a bow and a crown was given unto him." We see a bow but no arrows, yet it is also said of him "he went forth conquering and to conquer." Jesus is already crowned. He is King of kings and Lord of lords. How did this rider go forth conquering and to conquer without any arrows? The answer is:by his philosophies and religious teachings.

It is so easy to pick up on something that is not necessarily true if we are not very careful. One thinker put it this way. "That floating mass of thoughts, opinions, maxims, speculations, hopes, impulses, aims, aspirations at any time current in the world, which may be impossible to seize and accurately define, but which constitutes a most real and effective power, being the moral or immoral atmosphere which at every moment of our lives we inhale, again inevitably exhale, the subtle informing spirit of the world of men who are living alienated and apart from God." (Trench.)

Paul describes the ministry and teachings of the rider of the white horse of Revelation, the sixth chapter in his second letter to the Thessalonians. There were some of Paul's adversaries teaching that Jesus had already returned. They had written letters in Paul's name attesting to this untruth. Paul is attempting to correct this situation. He writes, "Now we beseech you, brethren, by the coming of our Lord Jesus Christ, and by our gathering together unto Him, That ye be not soon shaken in mind, neither by spirit, nor by word, nor by letter as from us, as that the day of Christ is at hand. Let no man deceive you by any means: for that day shall not come, except there be a falling away (apostasy) first, and that man of sin be revealed, the son of perdition; Who opposeth and exalteth himself above all that is called God, or that is worshipped: so that he as God sitteth in the temple of God, showing himself that he is God. (II Thes. 2:1-4.) This scene Paul is describing takes place during the last three and one-half years of the Great Tribulation Period. It demonstrates the fullness of the ministry of the rider of the white horse of Revelation, the sixth chapter.

The son of perdition, of course, is the antichrist. His revealing is already appointed by God. If Satan had his way, the antichrist would appear right now in person with all his false doctrines and lying signs and wonders but he must wait until his appointed time. The one thing that keeps him at bay is the presence of the Church. When the Church is taken out of the way, then he will come. His ministry will be very religious, lying signs and wonders and great delusions. Deception will come into play such as the world has never seen, nor ever will see again. The antichrist will have a church, an organization, a religious institution, but it will be founded on lies and deception and doctrines from hell. The giddy world will follow him and his deception. How horrible! What desolation! For three and one-half years, hell will be reigning. Now we can understand more fully the words of Jesus, "Your house is left unto you desolate." Paul said, "God shall send them strong delusion, that they might believe a lie." The reason? "Because they had not a love for the truth." (II Thes. 2:10,11.)

The 2nd seal is opened in Revelation, the sixth Chapter and a red horse appears. Its rider was "given the power to take peace from the earth, and that they should kill one another: and there was given unto him a great sword." (Rev.6:4.) We got a glimpse of this red horse and its rider on September 11, 2001 in the total destruction of the two towers of the World Trade Center in New York and partial destruction of the Pentagon in Washington, D.C. At this writing we understand that between 2,500 and 3,500 people perished in a matter of minutes. The color of the horse is red, symbolizing bloodshed. There is no doubt that this incident took peace from the earth. It is most definitely a prophetic event.

Just as the apostate revival began decades ago, acts of terrorism have also occurred for many years. The blowing-up of airliners, the bombing and destruction of government buildings such as in Oklahoma City a few years ago, the bombing of Embassies in different parts of the world, to mention a few. The attack on the World

Trade Center Towers was the greatest so far.

The "great sword" of the rider of the red horse of Revelation, Chapter 6 is "machaia." We get our English word "machete" from that Greek word. Now we know the machete is not great in size. It is rather small, ranging from perhaps 18 inches to 30 inches. It is called a "great sword" in the Bible because of the great disastrous work it does.

I know the four horsemen of Revelation, the sixth chapter, are symbolic of post-rapture Tribulation events, but they are, in a real sense, happening right now. In fact, they have been happening from time to time over the past few years, but are intensifying now. We can almost hear the prancing of the four horses on the perimeter of Tribulation events. The activity of these Tribulation events will be greatly intensified after the Rapture, the Catching Away of the Church.

The Scripture does not reveal to us when Jesus will return to this earth. No one knows. Jesus said that before His ascension. "But of that day and hour knoweth no man, no, not the angels of heaven, but my Father only." (Matt. 24:36). He gave us the signs to watch for that will take place just prior to His coming. We are instructed by Jesus to be ready. We must content ourselves with His instructions. To do more than this we place ouselves with the disciples as they questioned Him on the Mt. Of Olives just before He ascended. They asked Him, "Lord, is this the time when You will re-establish the kingdom and restore it to Israel?" He answered them, "It is not for you to become acquainted with and know what time brings—the things and events of time and their definite periods--fixed years and seasons (their critical nick of time), which the Father has appointed (fixed and reserved) by His own choice and authority and personal power." (Acts 1:6,7. Amplified Bible).

As far as we can see, there is nothing else that has to happen or Scripture to be fulfilled before His coming for His Bride. His coming may be at any time! It is EVER-IMMINENT!

In the first stage of the Second Coming of Jesus, no one will see Him. He comes "as a thief in the night." He quietly steals His Bride away. In the 2nd stage of His Second Coming, "Every eye shall see Him." We will discuss this in a later chapter.

Chapter Three

The Two Witnesses

And I will give power unto my two witnesses, and they shall prophesy a thousand two hundred and three score days, clothed in sackcloth. (Rev. 11:3. KJV)

The catching away of the Church in the Rapture will be heaven's signal for the seven year Tribulation Period to begin. Looking at the Book of Revelation from a chronological viewpoint, it will also signal the beginning of the ministry of the two witnesses. There are two time-periods mentioned in the beginning verses of Revelation, the 11th Chapter. One is mentioned in the above text. The two witnesses shall prophecy for "a thousand two hundred and three score days." This is forty-two months, or three and one-half years. This time-period is different from the one mentioned in verse 2. There, it is stated, "But the court that is without the temple leave out, and measure it not; for it is given unto the Gentiles: and the holy city shall they tread under foot forty and two months." It is safe to say that the ministry of the two witnesses will be the first half of the great tribulation period and the treading underfoot of the holy city by the Gentiles shall be the last three and one-half years of the great tribulation period. This will become more evident as we probe the depths of the doctrines of the last days.

As we have seen before, the Laodicean Age is the last Church Age of time as we know it. Its doctrines will be extended and intensified over into the Church doctrines prevalent in the Tribulation Period. Much could be said about such doctrines, but allow me to mention just a few things. There will be a noted intensification on the doctrines of prosperity within the church and outside the church. Political issues will be intensified. A system of democracy, of sorts, will be intensified and extended on an International level. Global politics will be greatly emphasized in importance during the Tribulation Period. An attempt will be made to stabilize social disorders, such as human rights, on a global basis. Great efforts will be made to bring the cultures of the East and West together and harmonize them. But the Bible says that all these things will not work to produce any lasting solution. The mingling of the politics of the East and West will not mix. Neither will the social disorders be remedied. The cultures cannot be harmonized. And most certainly, the religions of the East and West can never agree nor find any Scriptural stability. It is a matter of the strong trying to mix with the weak. This is the character of the kingdoms and governments of the last times; partly strong and partly weak. We are in those days right now , in the 21st century. Let me quote a passage of Scripture. "And whereas thou sawest the feet and toes, part of potters' clay and part of iron, the kingdom shall be divided; but there shall be in it of the strength of iron, forasmuch as thou sawest the iron mixed with miry clay, so the kingdom shall be partly strong and partly broken. And whereas thou sawest iron mixed with miry clay, they shall mingle themselves with the seed of men: but they shall not cleave one to another, even as iron is not mixed with clay." Daniel 2:41-43. The iron will of the dictator and the fickle mind of democratic thought will never mix. Nevertheless, it is such a world the two witnesses walk into to begin their last day ministry.

It will be a world of various political strategies with many social disorders; a world of clashing cultures and customs, and a world of different religions. Because of these things, compromise will be the order of international aims and goals in every aspect of the philosophies of the last times, religious as well as secular. Actually, peace will be much sought after but will be elusive. The Bible speaks of men crying, "Peace, peace; when there is no peace." Jer. 6:14.

In the time of the appearing of the two witnesses, the world will be very religious. In fact, there will be great efforts in play to synchronize all religions and bring them together. But the eastern mysticism will not mix with pure Christianity. Neither will Catholicism mix with Protestantism. There will be strong notions that mixing these doctrines is working but it will not be true. The character of the ministry of the two witnesses will be to project the Word of God as it is and as it always has been. Their ministry will be characterized by the miraculous works of power of God. The words, "They will be clothed with sackcloth," symbolizes their prayer life. They will have power to turn the rivers into blood and call fire down from heaven. They will have power to shut the heavens that it rain not in the days of their ministry.

They have power to "smite the earth with all plagues as often as they will."

I truly believe these two witnesses are Enoch and Elijah. They have never died and the Bible teaches that it is appointed to man once to die and then the judgment. These two men are men whom God took up to heaven alive. Enoch was taken up before the flood; Elijah during the days of the Jewish prophets.The Bible says Enoch "walked with God, and was not, for God took him." Gen. 5:24. That's not all the Bible says about him. Paul said in Hebrews 11:5 that he was translated that he should not see death. Jude said in the 14th verse of his Epistle that Enoch spoke of the evil of the last days, the false prophets and false believers that false prophets produce, "Saying, Behold, the Lord cometh with ten thousands of His saints, to execute judgment upon all..." So Enoch preached the second coming of Jesus as he was inspired by the Holy Ghost even in that ancient day before the flood.

Elijah was caught away to heaven by a whirlwind, but before his going to heaven, he had a powerful ministry in the Lord. He closed the heavens and no rain came for three and one-half years. He prayed fire down from heaven that consumed the sacrifice, the altar, and the dust around the altar. He prayed again and the Lord sent rain. In addition to all these miracles, he raised the dead and did many other miracles. Then he was caught up to heaven alive by a whirlwind.

These men will continue their ministry during the seven year Tribulation Period spoken of in the Book of Revelation. And the world will hate them. The church of antichrist believers will also hate them and they shall be killed. But their ministry will last three and one-half years. During this time period several attempts will be made on their lives but God will sustain them until the appointed time of their martyrdom. The forces of the antichrist mob will do the job. These are not just ordinary deaths. They are Providential and they have a purpose within the scope of the doctrine of the last days.

The two witnesses will be consumed with their mission. They know what they are about. They will not be golfing with the President, nor the Premier, nor the Prime Minister, or King. They will be men filled with the power of the grace of God. One thought consumes them, and that is to give God's final call to a lost and dying world. When they have "finished their testimony," they shall be killed by the Beast. Their deaths are predetermined and ordained by God Himself. They will be hated individuals. After they have been gunned down in the streets of Jerusalem, their dead bodies will lie in the streets of that city for three and one-half days. No one will be allowed to give them a decent burial. Yet, their ministry is not quite over.

How does the church of the Tribulation Period react to the death of these two men? Let the Scripture speak. "And they that dwell upon the earth shall rejoice over them, and make merry, and shall send gifts one to another; because these two prophets tormented them that dwelt on the earth." Rev. 11:10. The world has made its case. It has rejected their ministry and greatly antagonized them until finally they are slaughtered in the streets. To the world, it has rid itself of these two vagabonds who had nothing good to say about it. To the world they were unwanted vagabonds, but to God they were rich from heaven. To the world they were troublesome outsiders and strangers and rebels that would not fit into its mold, but to God they were citizens of heaven empowered with the message of grace in the last times. To the world they were dead men, their bodies decomposing as they lay in the streets of Jerusalem, but they will live again!! God is not through with them yet.

The Bible says, "And after three days and an half the spirit of life from God entered into them, and they stood upon their feet: and great fear fell upon them which saw them. And they heard a great voice from heaven saying unto them, Come up hither. And they ascended up to heaven in a cloud; and their enemies beheld them." Rev. 11:11, 12.

This is the end of the ministry of the two witnesses to the people of the Tribulation Period. Rev. 11:9 indicates that the entire world will see them as they lie on the streets of Jerusalem. It will be an International scene. It is the International Community that will not let them have decent burial. Before that, it is the International Community which is permeated with the spirit of antichrist that kills them. It is the power of the Beast (antichrist) that opposes them from the beginning of their last day ministries. Their death and resurrection mark the half way point in the seven year Tribulation Period.

The two witnesses were actually raptured. The eyes of the world are upon them as God resurrects them and catches them away. Revelation 11:9 reads as follows. "And they of the people and kindreds and tongues and nations shall see their dead bodies three days and an half, and shall not suffer their dead bodies to be put in graves." God has a purpose in this. He wants the world to see His power even at this late stage of time. The Bible students and scholars of past generations wondered at this verse of Scripture. But that was before television and the vast scope of our present day technology.

The martyrdom of the two witnesses will be the hottest news item on the globe. Every aspect of the last days society, both religious and political, will be gloating and rejoicing over their death. The fact that they will not be allowed a decent burial gives us the international mood concerning them. They will be the subject of every newscast, morning, noon and evening. ABC, CBS, NBC, CNN and every other News Organization will be there, each vying for the best vantage point. God's purpose for the two witnesses being there is not quite fulfilled. Let's view the scene as it is depicted in the Scriptures. They have been dead now for three and one-half days. They have been in full view of the eyes of the International Community. Their bodies are beginning to decompose, yet the world goes on with its rejoicing and celebrating. These two men who so often jammed the "peace process" with their narrow-minded ministries are now dead. They will no more torment the world with thir plagues. Their preaching and prophecies concerning a victorious Israel will no longer disturb the religious and political thought process. But God is not quite through with them yet. For, suddenly, the breath of life from God enters into them and they stand upon their feet. Not only will the world see these two men raptured, for that is what it will be, they will also hear a voice, a distinct voice and a mighty one saying, "Come up hither. And they ascended up to heaven in a cloud, and their enemies (the International Community) beheld them." (Rev. 11:12b) All of this will be captured on television cameras. Men will see it on their computers, for it will be Web-Cast on the Internet. What a scene it will be!What magnanimous proof that God is and His Word is sure! What infinite proof of the resurrection of Jesus Christ, Himself! These unburied, decomposing bodies are changed in a moment to glorified bodies and rise as their Lord had before them. What greater proof could God give that all is well in heaven and His word is sure and certain?

Chapter Four

The Two Great Wonders (Signs) in Heaven

And there appeared a great wonder in heaven; a woman clothed with the sun, and the moon was under her feet, and upon her head a crown of twelve stars: And she being with child cried, travailing in birth, and pained to be delivered. And there appeared another wonder in heaven; and behold a great red dragon, having seven heads and ten horns, and seven crowns upon his heads. And his tail drew the third part of the stars of heaven, and did cast them to the earth: and the dragon stood before the woman which was ready to be delivered, for to devour her child as soon as it was born. And she brought forth a man child, who was to rule all nations with a rod of iron: and her child was caught up unto God, and to his throne. And the woman fled into the wilderness, where she hath a place prepared of God, that they should feed her there a thousand two hundred and threescore days. (Rev. 12:1-6).

After the rapture of the two witnesses there was a great earthquake. A tenth part of the city fell and seven thousand men were slain. Verses 15 through 19 of the

11th Chapter of Revelation record the sayings of the voices in heaven as they praised God for magnifying His name on the earth. In summary of this passage, the raptured saints in heaven are thanking God for the consummation of all things. They view it as being ever-eminent. All things are ready! It is about to happen! But there are some things that must occur before their praises can come to fruition.

There are two signs shown to John in Chapter 12 of Revelation. "And there appeared a great wonder (sign) in heaven; a woman clothed with the sun, and the moon under her feet, and upon her head a (victor's) crown of twelve stars: and she being with child cried, travailing in birth, and pained to be delivered." (Verses 1,2). Some interpreters take this woman to be Mary and the child she is carrying to be Christ. While we are to respect their views and their efforts, let us interpret Scripture within the scope of its context. Nowhere in Scripture is Mary portrayed in such a manner. In truth, the woman is a sign or a type of the Organized Church. The twelve Tribes of Israel are very prominent and included in the full symbolic picture of the woman as the crown on her head. Let me hasten to say that while the Jews as a Nation have rejected Christ as the Messiah, the Jews can accept Him as individuals as did the Apostles in their day. They wrote of Him as the Only begotten Son of the Father. So, upon their conversion, the Jew of the 21st Century can be a part of the Christian Church.

When the woman's offspring was born, he was immediately "caught up unto God, and to His throne." Christ was not. He spent 33 years on the earth.

While the Organized Church has erred many times

in its practices and teachings; while it has shamefully compromised many times over the centuries, it has still contained the truth as it is revealed in the Scriptures, and from time to time real revival has broken forth when some divine truth was discovered and projected before the people.

The Church portrayed in the sign of the sun-clad woman can be viewed in two ways, as the visible Church and as the invisible Church. The visible Church is organized Christianity in general. Since its institution in the days of the Apostles, organized Christianity has possessed the truth and power of the Gospel of the Lord Jesus Christ. The truth has been nestled within the teachings of the Church even though it has been misconstrued and misunderstood by some. From these misunderstandings sometimes personalized convictions are taught instead of the truth. Thus, error creeps in. It has been so since the days of the Apostles. The Apostle John says, "Little children, it is the last time: and as ye have heard that antichrist shall come, even now are there many antichrists; whereby we know that it is the last time. They went out from us, but they were not of us; for if they had been of us, they would have no doubt remained with us: but they went out, that they might be made manifest that they were not all of us." (I John 2:18,19 KJV)

The truth is that organized Christianity has within it all the truth and the power of the grace of God. The truth also is that everyone that is a part of organized Christianity does not necessarily accept the truth and grace of God in its fullness. But there are some within the scope of the Organized Church who do believe with all their beings the Gospel in all its power. They are the purified ones, cleansed by the blood of their Lord. They are nestled within the body of the Organized Church or organized Christianity. They comprise the Invisible Church. We must not dare to attempt to name them. God, the righteous Judge knows them by name. They reside within the body of the Visible or Organized Church and comprise the "ready ones." They are the Bride of Christ. They are the offspring of the Sun-clad Woman. They are the ones that Satan would destroy if he could.

The "man child" of the Sun-clad Woman is neuter gender. "Nay, the letter of the description is such as to prove that this child is collective and composite, the same as the mother, and likewise includes people of both sexes. The word ARSEN, which means male, has the peculiarity of being the neuter gender, and so applies to both men and women, and cannot apply to any one individual." (J. A. Seiss, *The Apocalypse*. Page 298). This "man child" is also described as one "who was to rule all nations with a rod of iron." Jesus promises in Rev. 2:26, "And he that overcometh, and keepeth my works unto the end, to him I will give power over the nations: And he (the overcomer) shall rule them with a rod of iron; as the vessels of a potter they shall be broken to shivers: even as I received of my Father." So the offspring of the Sun-clad Woman is the true believer or the Invisible Church that is raptured "caught up" to heaven and to God's throne. In his Expanded Translation, Kenneth S. Weust translates it like this: "And the child was snatched up to God and to His throne."

The second wonder (sign) that John saw in Rev. 12

was a great red dragon with seven heads and ten horns. He was standing before the woman ready to devour her child as soon as it was born. The Bible is very explicit in its description of the dragon. "Upon his heads were seven royal crowns." (The Expanded Translation, Kenneth S. Weust). He is also identified as "that old (ancient) serpent, called the Devil, and Satan, that deceiveth the whole world (the whole inhabited earth). (Rev. 12:9).

The description of the crowns upon the dragon's heads is very significant. We must go back to the beginning in the Garden of Eden to get the full picture. God had made Adam king of the earth and given him the responsibility of filling the earth with people just like him. Of course, at that time he had not sinned. So the command was to fill the earth with a holy people. Before Adam could begin to carry out the command of God, Satan in the form of the serpent tempted and deceived Eve and Adam. Adam forfeited his Kingship and turned it over to Satan through deception. Satan mightily influenced the antediluvian generations in their rebellion against God. So the flood came and "destroyed them all," except Noah and his family. As the Godly lineage was carried on through Seth and his descendants before the flood, it was carried on through Shem and Japheth after the flood.

It was Ham's rebellion against Noah and his God that continued the struggle between Godliness and ungodliness.Ham's domain in that ancient day was Egypt. (Psalm 105:23). Ham's grandson, Nimrod, continued the rebellion against God. His domain was Assyria. These were the first two World Empires, if we can call them such, after the flood. Their leaders, Ham and Nimrod, were ingenious men , especially Nimrod. Satan used their ingenuity in his rebellion against God. Following these two world Kingdoms came Babylon under Nebuchadnezzar and his dynasty, the Medo-Persian Kingdom under Cyrus and Darius, the Grecian Kingdom under Alexander, the Great, and Rome under the Caesars.There is one more world Kingdom that now is in the making. If we total them up, there are seven. These are the seven crowns on the heads of the Great Red Dragon of Rev. 12. He has in a real sense headed up the World System in its seven historic periods since the flood.

Now Satan is a spiritual creature. He is evil and is called the Prince or Ruler of devils in the Scripture. He was once known as "the Son of the Morning" in heaven, but he rebelled against God and attempted to overthrow Him and was cast out of heaven, influencing and taking a third part of the angels with him. His domain has been the atmosphere around the earth since that time. In God's appointed time he will be cast down to the earth by Michael and his warring angels. Then, at that time, he will be limited to the geographical earth. This will be the beginning of the last three and one-half years of the Tribulation Period.

The Sun-clad Woman and the Great Red Dragon of Revelation,the 12th Chapter, portray the struggle and spiritual warfare that has been going on since the Church came into existence. Satan would destroy it if he could. But thank God, Satan can only do what God allows him to do. Satan cannot curse the Church because it is blessed. It is the dwelling place of God and it is victorious.

God has other things appointed for this old world and they shall come to pass in their appointed time. Hear God's Word as it speaks. "And I heard a loud voice saying in heaven. Now is come salvation, and strength, and the kingdom of our God, and the power of His Christ: for the accuser of our brethren is cast down, which accused them before our God day and night. And they overcame him by the blood of the Lamb, and by the word of their testimony; and they loved not their lives unto the death. Therefore rejoice, ye heavens, and ye that dwell in them. Woe to the inhabiters of the earth and the seas! For the devil is come down unto you, having great wrath, because he knoweth he hath but a short time." (Rev. 12:10-12).

Chapter Five

The Beast from the Sea

And I saw a Wild Beast coming up out of the sea, having ten horns and seven heads, and upon his horns ten royal crowns and upon his heads names, the essence of which is impious and reproachful speech injurious to the divine majesty of Deity. And the Wild Beast whom I saw resembled a leopard. And his feet were like those of a bear. And his mouth resembled a lion's mouth. And the Dragon gave him his [the Dragon's] miraculous power and his throne and great authority.

And one of his heads appeared to have been mortally wounded, the throat having been slashed. And his death stroke was healed. And the whole earth followed after the Wild Beast in amazement. And they worshipped the Dragon because he gave the authority to the Wild Beast. And they worshipped the Wild Beast, saying, Who is like the Wild Beast, and who is able to go to war with him?

And there was given to him a mouth speaking great things and things injurious and reproachful to the divine majesty of Deity. And there was given to him authority to operate forty-two months. And he opened his mouth in a slanderous attack against God, to His face, to defame His name and His dwelling place, [and] those who dwell in heaven.

And there was given to him [permission] to make war with the saints and to gain the victory over them. And there was given to him authority over every tribe and people and language and nation. (Rev. 13:1-7. An Expanded Translation, Kenneth S. Wuest.)

We must understand that the devil does not always appear as a Great Red Dragon. In fact, he is more likely to appear as an "angel of light." Let me quote what the inspired Apostle Paul wrote to the Corinthian Church. "For such are false apostles, deceitful workers, transforming themselves into the apostles of Christ. And no marvel; for Satan himself is transformed into an angel of light." (II Cor. 11:13, 14.) Satan was seen by John, the inspired seer, in his true character; a horrible creature of killings, and blood, and deceit, and lies. He is the epitomy of evil, the master of deceit. There is no good thing in him. The description of the Great Red Dragon, as John saw him, is symbolic of all this.

There is a divinely appointed time for Satan to be cast down to the earth, and it will occur at that precise time; not a minute sooner and not a minute later. Satan knows this! Satan may not know when that moment will come, but God knows. When that moment comes and Satan realizes that he is cast down to the literal earth, the Scripture says that he has "great wrath, because he knoweth that he has but a short time." (Rev. 12: 12.) For this reason, he goes forth to persecute the remnant of the Church that is left on the earth after the rapture of the ready ones.

I feel we must identify those who comprise the remnant. They are the people who have sat under the teaching of the Scriptures and at some time in their lives have seen and understood the truth and for some reason have never surrendered themselves to those divine truths, or they could be those who once followed the truths and for some reason gave up and went back to their former way of living. In either case, when the rapture occurs, they will understand it for what it is and know that they have been left behind. Many of them will at that time surrender themselves to God and will suffer this cruel persecution of the devil.

The Great Red Dragon goes to the shore of the sea and calls forth another beast who greatly resembles him. This beast has seven heads and ten horns just the same as the Great Red Dragon. But there is a difference. In the case of the Great Red Dragon, his heads were crowned with royal crowns. In the case of the second beast from the sea, his horns are crowned with royal crowns. There is a reason for this. The devil has ruled over the seven world powers that have risen since the flood in the days of Noah. His son, the second beast who is the antichrist, will rule over ten kingdoms during the last three and one-half years of the Great Tribulation Period, or, at least in the beginning of his reign. Early in his reign, he will destroy three of the kingdoms, leaving seven remaining.

You see, the work of the seventh world leader is to divide the geographical earth into ten divisions which will constitute the last ten kingdoms. This is in formation now, at the time of this writing. More will be said about this later and Scripture given for support.

We must note that the second beast is the Antichrist and his character is symbolized as a beast, just as Satan's character is symbolized as the Great Red Dragon. Let me repeat that John, the seer of divine things, saw both the Great Red Dragon and the Antichrist, in their true character, not as the world will see them. While the Antichrist will rule over the entire world, divided into ten kingdoms, symbolized by the ten horns, we must not forget his seven heads. They symbolize the same here as they did in the case of the Great Red Dragon. And that is the seven World Empires since the flood. Seven World Kingdoms and seven World Rulers or Kings are represented here. One of them was murdered. His throat was slashed and he was dead. But he shall live again. This is the Antichrist. This is the Wild Beast called up out of the sea. This is not a sea of water. It is the sea of humanity. The Beast is a man who has once lived, and who was killed and went to hell or the bottomless pit. He is resurrected from this dreadful place to live again. How do we know this is true? Because the Bible describes him as "the beast that was, and was not, and yet is." (Rev. 17:8). This Ruler or King will be resurrected. He will be called forth from hell at the appointed time and he will rule again. His dwelling place at this writing is the bottomless pit. He is imprisoned there and cannot come forth until God is ready, but he will come in his time! This is Bible prophecy!

So the Antichrist is a master of deception. His mission and purpose will be to convince the world of humanity, including the religious sector, that he is the expected Messiah. God allows Satan to call him forth. God allows Satan to give him his power and authority. The word

"Antichrist" has two meanings. He is against Christ, and he is a substitute Christ. He also is the son of perdition, the son of hell. His appearance greatly resembles his father, the devil. He has seven heads and ten horns, as does his father. Satan is a spirit being. Insomuch as the beast has been raised from the dead, he, also, is a spirit being. He will not only work religious miracles but will also have the ability to appear and disappear at will. His deception will be that complete. Thus far, only Jesus has had that ability. This ability, along with his "signs and lying wonders," (II Thes. 2:9) will be enough to convince a fickle and frivolous world that he is the Messiah who should come.

Some Bible scholars and students of prophecy believe that the Antichrist will come as a man of flesh, bone, and blood and then be killed and only then to rise from the dead. People sometimes ask the question, "Do you think the Antichrist has already been born?" Let me emphatically answer that question. Yes!! He has been born and probably already killed and waiting in the bottomless pit to make his appearing at the time appointed by God. This most certainly is true unless the seventh world ruler, who hasn't been appointed yet, happens to be this awful personality. Personally, I believe the Antichrist will be Nimrod, son of Cush, and grandson of Ham, Noah's son. After all, he is the progenitor of the world system which has always been alienated from God.

What will the World System be like in the last days? To answer this question we must consider at least two aspects. "Kosmos, or the Organized World System can be viewed in two different ways. (1) When used in the New

Testament of humanity,, the world of men, it is organized humanity---humanity in families, tribes, and nations---which is meant. The word for chaotic, unorganized humanity---the mere mass of men---is thalassa, the "sea" of men." (C.I. Scofield's note on Matt. 4:8.)
(2) "In the sense of the present world system, the ethically bad sense of the word refers to the "order", "arrangement" under which Satan has organized the world of unbelieving mankind upon his cosmic principle of force, greed, selfishness, ambition, and pleasure. This world system is imposing and powerful with armies and fleets; is often outwardly religious, scientific, cultured, and elegant; but seething with national commercial rivalries and ambitions, is upheld in any real crisis only by armed force and is dominated by Satanic principles." (C.I. Scofield's note on Rev. 13:8.)

There is at the present time a mysterious and magnetic pulling of the World System together. This is true in every aspect of the world's societies. Soon there will be a One-World Government, a One-World Religion, and an attempted One-World Cultural Society, plus a One-World Language. All of these are steadily in progress right now. Can anyone deny the fact that there is a concerted effort right now to spread democracy, at least in some form, over all the world? Democracy means "ruled by the people." In a democratic society, leaders are elected by a majority of votes cast by the people. Already elections decided by popular vote have taken place in many nations. When the Democratic Process grows to include the International Community, does this mean that Nations will lose their sovereignty? Does this mean that the United States of America, for instance, will have to dissolve its Congress and Constitution, and not have elected Officials, and a President and Cabinet? No, it doesn't mean that, at all. Each Nation could continue much as it is now. However, World Courts will be much more prominent and will be expected to abide by the Rule of International Law. It has already been referred to as "the rule of law", in a "New World Order." In a One-World Government scenario, national officials will lead in a vassal capacity under an International Federal Head of State.

How about a One-World Religion? It will come about by compromise of the cardinal doctrines of the Protestants, Catholics, Jews, and Muslims and any other religion or belief system in existence. The pretense of unity will be greatly emphasized in bringing about such a concept. Consequently, the One-World Religion will not be a pure religion.

The cultures of the world can never mix. It will be only an attempt. In Daniel's interpretation of Nebuchadnezzar's dream, there were four Kingdoms mentioned. The head of gold was the Babylonian Kingdom, the breast and arms of silver was the Kingdom of the Medes and Persians, the belly and thighs of brass was the Grecian Kingdom under Alexander, and the legs of iron and feet and toes of part iron and part clay was the Roman Kingdom. The ten toes represent the last Ten Kingdoms that will exist under the rule of the Antichrist. Symbolically, they are the same as the ten horns of the Wild Beast which is the Antichrist. Pure religion will not mix with impure religion, or impurities in any form. Neither will the cultures of the East and West mix. The Holy Spirit through Daniel put it this way, "And whereas thou sawest iron mixed with miry clay, they shall mingle themselves with the seed of men: but they will not cleave one to another, even as iron is not mixed with clay." (Dan. 2:43.)

As for the One-World Language, it will be English. As a result of the conquests of Alexander, the Great, the Greek language became universal. So will the English language be universal in the end time.

The foregoing concepts only partially describe the times of the Wild Beast and Antichrist regime. There are more. There will be a One-World Currency. This has already been put in place in the European Community and is working fine. It enhances the trade market of those countries involved in the EC, as well as travel and other aspects of interchange. The Bible says that the Antichrist shall rule over all the world, every language, every tribe, and every nation. In fact, Scripture teaches that the world of unsaved mankind will worship him.

In the 17th Chapter of Revelation, we see a woman sitting on a Beast which has seven heads and ten horns. This is the same Beast we saw in the 13th Chapter of Revelation. He has the same appearance. The seven heads mean the same and so do the ten horns. Here, every thing about the Beast and the woman is explained. There are certain things we must recognize in order to get the proper interpretation and meaning that is depicted here. The woman, the great prostitute, is sitting upon many waters and she is about to be judged. (V. 1) In the interpretation given by the angel that showed John these things, it is stated, "The waters which thou sawest, where the whore sitteth, are peoples, and multitudes, and nations, and tongues." (V. 15) The woman's identity is ever-plain in the Scripture. God has forever identified her for us. "And upon her forehead was a name written, MYSTERY, BABYLON THE GREAT, THE MOTHER OF HARLOTS AND ABOMINATIONS OF THE EARTH." (V 5) Notice how the woman is dressed and how she stands out in her appearance. "The woman was arrayed in purple and scarlet colour, and decked with gold and precious stones and pearls, having a golden cup in her hand full of abominations and filthiness of her fornication. (V. 4) She is an attractive personage and symbolizes everything the World System has to offer. Her beauty attracts men, both small and great. She especially entices Kings, and Premiers, and Presidents and World-Leaders of all titles and descriptions. She, most certainly, symbolizes World Trade and International Commerce. She has lived and reigned for many centuries. In fact , she says of herself, "I sit a queen, and am no widow, and shall see no sorrow." Rev. 18:7b.)

Through prophetic eyes, the inspired prophet, Nahum, sees her in his day and describes her thus, "Because of the multitude of the whoredoms of the well-favoured harlot, the mistress of witchcrafts, that selleth nations through her whoredoms, and families through her witchcrafts." (Nahum 3:4.) The New Jerusalem Version describes her in this verse as "the graceful beauty, the cunning witch." She has promised gratification of every desire known to humanity and allured both kings and paupers alike. And she is carried by the Beast, the personification of deceit and false religions of every age.

When John saw this woman sitting upon the many waters and carried by the Beast, he marvelled or geatly wondered.. "The angel said unto me, Wherefore didst thou marvel? I will tell thee the mystery of the woman, and of the beast that carrieth her, which hath the seven heads and ten horns." (Rev. 17:7.) The Beast is described in V. 8 as one who ascends out of the bottomless pit and goes into perdition. All the people of the earth who are not saved will wonder when they see him. He is described as one who was and was not and yet is, meaning he once lived, and died, and yet here he is living again. He is resurrected from the dead. His seven heads are seven mountains or Kingdoms. David once thanked the Lord for his Kingdom by saying, "Lord, by thy favour thou hast made my mountain, (meaning his kingdom.) Psalm 30:7. The mountains mentioned here in V. 9 do not refer to the seven mountains of Rome as some interpreters think. They are seven Kingdoms on which the woman sits. The Kingdoms each have a king. Five of these kingdoms had already existed before John's time. They were: (1) The Ancient Egyptian Kingdom, (2) The Assyrian Kingdom, (3) The Babylonian Kingdom, (4) The Medo-Persian Kingdom, and (5) The Grecian Kingdom. These five had risen and fallen. "One is" was The Roman Kingdom which existed in John's time, and was the 6th Kingdom. "The other that is not yet come" is in the making right now, in the 21st century. Its leader has not yet been elected, but when he comes into power "he must continue for a short space." (V. 10) The seventh king or world leader will die, probably through assassination. This means that all seven Kings have died. One of the seven shall be resurrected, (not necessarily the last) and live again in his resurrected body. This is Antichrist! "And the beast that was, and is not, even he is the eighth, and is of the seven." V. 11. This is Antichrist!

The ten horns of the Beast are the last ten Kingdoms that constitute the International Community of the last days. They are comprised of the mixed cultures and religions of the world and will not cleave one to another.

"And the woman which thou sawest is that great city (or System) which reigneth over the kings of the earth." (V. 18.) Her name is Babylon! Her name is Confusion!

Chapter Six

The Beast out of the Earth

And I beheld another beast coming up out of the earth; and he had two horns like a lamb, and spake as a dragon. And he exerciseth all the power of the first beast before him, and causeth the earth and them which dwell therein to worship the first beast, whose deadly wound was healed. And he doeth great wonders, so that he maketh fire come down from heaven on the earth in the sight of men. And deceiveth them that dwell on the earth by the means of those miracles which he had power to do in the sight of the beast; Saying to them that dwell on the earth, that they should make an image to the beast, which had the wound by a sword, and did live. (Revelation 13:11-14. KJV)

There are two Beasts mentioned in the 13th Chapter of Revelation. One arose out of the sea of humanity. The other arose out of the earth. The first was one who had lived before. The second had not. The first Beast was resurrected from the bottomless pit to live again on the earth as the "son of perdition." The second beast is a man of the earth, a man of flesh, bone and blood. The first Beast is the prophetic Antichrist. The second Beast is his prophet. It is this False Prophet that we want to discuss now.

You'll notice that the second Beast appears as a lamb with two horns, but he speaks as a (the) Dragon. He looks very religious, but his message is the message of the Devil. "And he exerciseth all the power (authority) of the first beast before him, and causeth the earth and them which dwell therein to worship the first beast whose deadly wound was healed." (Rev. 13:12.) It is said of the first Beast, "and the dragon gave him his power, and his seat (throne), and great authority." (Rev. 13:2b.) This is, indeed, the diabolical trinity from hell: Satan, being the "god (prince) of this world (world-system)" as Jesus said in John 14:30, the Antichrist being his son, the "son of perdition," (II Thes. 2:3), and the False Prophet being the messenger of this incredibly evil trinity.

The fact that the False Prophet makes his appearance as a lamb is nothing new. The Bible says this is the plight of all false prophets. "For such are false apostles, deceitful workers, transforming themselves into the apostles of Christ. And no marvel; for Satan himself is transformed into an angel of light (minister of the Gospel). Therefore it is no great thing if his ministers also be transformed as ministers of righteousness; whose end shall be according to their works." (II Cor. 11:13-15). This Scripture will suffice in describing the false prophet of any time period.

The False Prophet of Revelation, Chapter 13 personifies the false prophets of all ages. He will be a preacher of preachers. He will look like a man of God and his mannerisms and demeanor will not betray this. He will be ultra persuasive in his method and ministry.

When Daniel saw the Beast with ten horns (kings or kingdoms) in his prophetic visions, he also saw a "little horn" (the Antichrist) come up in their midst. He was amazed at the great words that came out of his mouth. We must remember that the mission of the False Prophet is to parrot the message of the Antichrist and convince the generation of the last days that he is Jesus Christ visibly returned. How will he do that? Simply by vocalizing the same message as the Antichrist, a mixed message of some truth and a lot of error which is indicative of the "falling away" of II Thes. 2:3.

Peter uses an interesting word in I Peter 4:11. "If any man speak, let him speak as the oracles of God." The Greek word "oracles" was used in classical Greek as the oracular utterances of heathen deities. As Peter used it, it refers to divine utterances and revelations. To say the least, the False Prophet will put forth the utterances of Satan and his son, the Antichrist.

In addition to the persuasive rhetoric of the False Prophet, he will have miraculous powers, also. "And he doeth great wonders, so that he maketh fire come down from heaven on the earth in the sight of men. And he deceiveth them that dwell on the earth by means of those miracles which he had power to do in the sight of the beast." (Rev. 13:13, 14)

Let us look at the word "prophecy", and discuss what it means. The Bible tells us that "all Scripture is given by inspiration of God, and is profitable for doctrine, for reproof, for correction, for instruction in righteousness: that the man of God might be perfect, throughly furnished unto all good works." (II Tim. 3:16, 17.) Also, Peter tells us in explaining the true sacred measure of prophecy, "Moreover I will endeavour that ye may be able after my decease to have these things always in remembrance. For we have not followed cunningly devised fables, when we made known unto you the power and coming of our Lord Jesus Christ, but were eyewitnesses of his majesty. For he received from God the Father honor and glory, when there came such a voice to him from the excellent glory, This is my beloved Son, in whom I am well pleased. And this voice which came from heaven we heard, when we were with him in the holy mount. We have also a more sure word of prophecy; whereunto ye do well that ye take heed , as unto a light that shineth in a dark place, until the day dawn, and the day star arise in your hearts: Knowing this first, that no prophecy of the Scripture is of any private interpretation. For the prophecy came not in old time by the will of man: but holy men of God spake as they were moved by the holy Ghost." (II Peter 1:15-21.) What Peter is really saying here is that the Scripture takes precedence over voices, visions, or any other thing; even if it seems to come from heaven. The idea is not that the holy men of old were "moved on." But they were moved by the Holy Ghost. This means that the Holy Ghost in some way moved them into the very presence of God where they heard the utterances of God and spake them. What a place in which to be! We could say, without stretching the meaning, that the holy men of old stood in the very utterances of God themselves. Therefore, we can boldly say that what God has stated in prophecy will most certainly come to pass.

Let's think on false prophets for a little. The false prophet is not of God. They are mentioned throughout the Bible. And they have prophesied lies throughout the Bible. They have been led astray by their own unscriptural aspirations or by the persuasion of some evil spirit. They usually are not seeking to please God but to please themselves. Let us hear what God has to say about them. "Thus saith the Lord of hosts, Hearken not unto the words of the prophets that prophesy unto you: they make you vain: they speak a vision of their heart, and not out of the mouth of the Lord. They say unto them that despise me, The Lord hath said, Ye shall have peace; and they say unto every one that walketh after the imagination of his own heart, No evil shall come upon you. For who hath stood in the counsel of the Lord, and hath perceived and heard His word? Who hath marked His word and heard it? Behold a whirlwind of the Lord is gone forth in fury, even a grievous whirlwind: it shall fall grievously upon the head of the wicked. The anger of the Lord shall not return, until He have executed, and till He have performed the thoughts of His heart: in the latter days ye shall consider it perfectly. I have not sent these prophets, yet they ran. I have not spoken to them, yet they prophesied. But if they had stood in my counsel, and had caused my people to hear my words, then they should have turned them from their evil way, and from the evil of their doings." (Jeremiah 23:16-22.) "Then the Lord said unto me, The prophets prophesy lies in my name; I sent them not, neither have I commanded them, neither spake unto them: they prophesy unto you a false vision and devination, and a thing of nought, and the deceit of their heart." (Jeremiah 14:14.)

We can see from the foregoing the character and

nature of all false prophets. The False Prophet, the Beast from the earth, of Revelation, Chapter 13 is the culmination of all false prophets. He gets his message and utterances from the first Beast, the Antichrist. He is the oracle and utterance of all evil of all time. The character and nature of the Devil, the Antichrist and all false prophets are personified in him. Persuasive? Yes! Alluring? Yes! Even Daniel was amazed at the great words he was saying as he saw him in his visions. (Dan. 7:11.) How much more the people who will be living in the time of the Tribulation Period?

The False Prophet is an individual entity of the last times whose purpose of being is to mislead the world by his incredible deceit. Because of his miracles he will be able to gain the religious trust of the people of his time. The bible says, "And deceiveth them that dwell on the earth by the means of those miracles which he had power to do in the sight of the Beast (Antichrist)". The message and works of all false prophets of all time will be intensified and symbolized in him.

Chapter Seven

The Ten Kingdoms

And whereas thou sawest the feet and toes, part of potters' clay and part of iron, the kingdom will be divided; but there shall be in it of the strength of the iron, forasmuch as thou sawest the iron mixed with miry clay. And as the toes of the feet were part of iron, and part of clay, so the kingdom shall be partly strong, and parrtly broken. And whereas thou sawest iron mixed with miry clay, they shall mingle themselves with the seed of men but they shall not cleave one to another, even as iron is not mixed with clay. (Daniel 2:41-43. KJV)

Since it is the business of the Seventh World Power Era to establish the boundaries of the Ten Kingdoms of the last days, it will be well to expand on this subject at this time. You remember, it was mentioned briefly in the fourth Chapter of this Book. The leader of this New World Order, which it is beginning to be called , "must continue for a short space" according to Revelation 17:10. Indeed, his main work will be to set the Ten Kingdoms in place. It will involve many areas of the International Scene and each area will have a significant Global impact.

According to the Holy Scriptures, there will be a mixture of Global politics with an emphasis on the Democratic persuasion. Also, an attempted mixture of the cultures of the east and west will come into focus. In fact, there will be an attempted mixture of the thought processes and belief systems of the Eastern and Western worlds. Every facet of living will be included. Can you imagine? A mixture of world politics, world cultures, and world religions with all their facets of teachings and doctrines will be pushed by a strong international demand for the co-operation of all people everywhere.

I, personally, have seen this in the Scriptures for many years, but I have honestly wondered how it would all take place. Would the people and rulers of the dictatorial Eastern nations finally see the advantage of the Democratic process and willingly convert to its teachings? How would the fulfillment of this Bible prophecy be brought about? Would there be a great religious revival that would birth such a mixture of the lifestyles of the Eastern and Western peoples? I think not!

I am writing this shortly after the so-called liberation of the Iraqi people. We have seen a war fought and won in just 27 days. Now, Iraq must be rebuilt western-world style. The brutal and dictatorial regime of Sadam Hussein is ended. Now the era of free speech, free will and free thought can begin. And it will seemingly work, at least for a while. You see, this is the beginning of the Ten-Toed Kingdom. Does the Scriptures teach the blending of the East and West is possible? Maybe! At least for a short time. "They shall mingle themselves with the seed of men but they shall not cleave one to another." Dan. 2:43. We are living in the last days and things and events are happening quickly. Prophecy is being fulfilled on a daily basis. The Ten Kingdoms must be set in place and that speedily.

We're living in the Laodicean Age, both secularly and religiously. The Democratic Process is very active in both areas. The majority rules. At least, that is the teachings of the Democratic process. It seems only right that Democracy and its freedoms should be extended to all peoples of the world. In fact, a good portion of the world of this day have adopted the Democratic process in some measure. The people of these nations are being allowed to vote for political leaders and some political processes. But the dictatorial thought still permeates the political structures in many of these countries. Let it be noted; the iron will of the dictator and the fickle will of democratic thought will never mix, at least for any meaningful period of time.

The Ten Toes of the image of Nebuchadnezzar's dream and the Ten Horns of the Antichrist are the same. They both portray the Ten Kingdoms of the last days. And it is true, we can see them beginning to be set in place now, in our time. What does it all mean? Or, perhaps we should ask another question. How near are we to the Second Coming of our Lord and Saviour, Jesus Christ? The light of Scriptural truth is shining all around us. The signs of the times are flashing like neon lights from every direction.

Bible prophecy is being fulfilled every day. This greatest sign shall take place suddenly, "For this we say unto you by the word of the Lord, that we which are alive and remain unto the coming of the Lord shall not prevent them which are asleep. For the Lord Himself shall descend from heaven with a shout, with the voice of the archangel, and with the trump of God: and the dead in Christ shall rise first: Then we which are alive and remain shall be caught up together with them in the clouds, to meet the Lord in the air: and so shall we ever be with the Lord." I Thes. 4:15-17. Dear reader, this will be the Rapture of the Church, the Bride of the Lord Jesus Christ. It is prophecy and it shall happen! After the Rapture, the Great Tribulation Period will set in. Things will not be as they are now.

The Ten Kingdoms of the last days are now being organized and are almost set in place. The Antichrist will make his appearance three and one-half years after the Rapture and install a World System such as the world has never thought of before. It is the doings of mankind that shall bring all these things upon humanity. These times will be the result of the political, religious, and social structures of the last days.

Mankind, on a global scale, is largely without God. It does not take Him into consideration when forming the strategies of world governments. Therefore, we can philosophically say that man without God is master of his own destiny. The results of his decisions are not naturally predictable with any certainty. It is, with this concept, we can begin to understand the meaning of true Bible prophecy. The Bible says, "God is not the Author of confusion, but of peace." I Cor. 14:33. So, it is the selfish nature of mankind that brings about the adverse circumstances of men as individuals and nations. God, in His infinite foreknowledge knew all this, so He moved His prophets of old to prophesy of future events. It is with divine certainty that God will intervene in the affairs of men and bring His Word to pass at the appointed times. Praise the Lord!

In the meantime, the world stage will continue to be set by men who are largely without God. "For nation shall rise against nation, and kingdom against kingdom: and there shall be famines, and pestilences, and earthquakes, in divers places. All these are the beginning of sorrows." Matt. 24:7, 8. These men who cause these things are world leaders and their intentions may be intended for the good of all mankind. God is allowing them to make their decisions through their free-moral-agency which He has given them. Since they have left God out of their reasoning and choices, He turns them over to a reprobate mind, and many times their decisions are not for the good of all mankind as they supposed. Dictators rise up to rule with iron-will authority. Cruel World Governments are put in place that suppress and stamp out the God-given freedoms and rights of humanity.

The establishment of the Ten Kingdoms of the last days will be man's major effort to establish some form of order that will be for the good of the International Community. It will be an effort to bring all nations together with little or no divisive factors. Wouldn't it be wonderful if man could achieve this utopia? Imagine, if you can, a one-world government, a one-world currency or monetary system, a one-world language, a one-world religion, and a one-world ruler! A concerted effort is being made now for such a System to be set in place.

A New World Order with a Rule of Law is in the making. There will be one Federal Head of this New World Order. Does this mean that nations will cease to exist as nations? No! This leader will allow them to govern their countries with vassal authority. Each nation will be allowed to continue with its elected President or Premier, and with its Congress or Parliament or whatever its legislative body may be called, but they will be subject to the International Rule of Law and an International Constitution administered by an International Court. After all, a World-wide Democracy is presently being established, isn't it? Will it work? Yes, or it will seem to work for a while. You see, this is the exact environment that the Antichrist will step into when he makes his appearance at his appointed time. "And the ten horns which thou sawest are ten kings, which have received no kingdom as yet; but receive power as kings one hour with the beast. These have one mind, and shall give their power and strength unto the beast." Rev. 17:12, 13.

The reign of Antichrist on Earth will be a terrible Regime. All the brutality and cruelty of all past world dictators will be combined into one personality. The Antichrist is "son of perdition," so named by Christ Himself. He is the son of Satan, straight from the abyss, the bottomless pit. While he is claiming to be the Messiah, there is no love, or holiness, or grace, or compassion about him at all.. He has the intelligence of Satan! What he doesn't know about the diabolical scheme of hell, Satan will teach him! His reign will be democratically religious, and seem to have the greatest concern for human rights, but in reality, it will be cruel, smashing and putting down everything and everybody that is not in harmony with the iron will of his dictatorship.

Everyone who takes the mark of his government,

666, will carry identification proving their allegiance to his government. The promise of prosperity and peace will be theirs! But in yielding to his mark, they, in reality, become citizens of hell, and participants in Satan's hellish scheme to overthrow God and occupy His throne for himself.

Dear reader, I want to ask you a question. If the Ten Kingdoms are, indeed, being set in preparation for the divinely appointed time of the coming of this horrible beast, how close is our generation to the Rapture of the Christian Church, which will be the divine cue for the beginning of the Great Tribulation Period and the Mark Of The Beast? The Bible says, "And I heard a loud voice saying in heaven, Now is come salvation, and strength, and the kingdom of our God, and the power of His Christ: for the accuser of our brethren is cast down, which accused them before our God day and night. And they overcame him by the blood of the Lamb, and by the word of their testimony; and they loved not their lives unto the death. Therefore rejoice, ye heavens, and ye that dwell in them. Woe to the inhabiters of the earth and of the sea! For the devil is come down unto you, having great wrath, because he knoweth that he hath but a short time." Rev. 12:10-12.

Chapter Eight

A Place Called Armageddon

And he gathered them together into a place called in the Hebrew tongue Armageddon. (Rev. 16:16. KJV)

Armageddon takes place at the end of The Great Tribulation Period. There will be a lot of Providential events that bring the human race to this place in prophecy. We know that Bible prophecy is sure and certain. It would be good at this point to mention a few things of interest that prove the certainty and accuracy of the Scriptures.

Let us hear what Daniel has to say on this subject. He had been taken captive during Nebuchadnezzar's reign. God had greatly used him in giving him dreams and visions and interpreting what God was saying to his generation and also to generations to come. He says, "I, Daniel, understood by books the number of the years, whereof the word of the lord came to Jeremiah the prophet, that He would accomplish seventy years in the desolation of Jerusalem." (Dan. 9:2.) When Daniel understood this, he began to pray, confessing his sins and the sins of Israel. He wanted to know what God was going to do and what the future held for God's people. As

Daniel prayed in dead earnestness, "O Lord, hear, O Lord, forgive, O Lord, hearken and do; defer not for thine own sake, O my God: for thy city and thy people are called by thy name." (V. 19.) While these words were still in his mouth, as Daniel was speaking to God, Gabriel appeared and began giving him an answer to his prayer.

Gabriel referred to another time in the near future of Israel. Cyrus, king of Persia would give a commandment to rebuild the walls of Jerusalem and the Temple. From that point in history there would be seventy weeks of years, or 490 years. These seventy weeks of years is what we want to look at now. Beginning with the decree of King Cyrus, there would be two periods of time: one would be seven weeks of years, which is 49 years, and the other would be sixty-two weeks of years, which is 434 years. Now if we add the 49 years and the 434 years together, the total is 483 years. The Bible says that at that time "Messiah shall be cut off," or crucified. (Dan. 9:26.) So, from the time that King Cyrus gave his decree that Jerusalem and the Temple should be rebuilt until the Cross was actually 483 years.That's 7 years short of the 490 years given by Gabriel. Let's stop right here and look at this time period in another way. We must start with the Babylonian captivity of the Jews.

The Jews were taken captive in three separate segments; one in the seventh year of Nebuchadnezzar's reign, again in the fourteenth , and finally in the twenty-third year of his reign. (Please refer to Jeremiah 52:28-30.)The final captivity occurred in 586 B.C, Now, let's do a little more arithmetic. If we subtract the 70 years of captivity from 586 B.C., we come to 516 B.C. If we again subtract 483 years from 516, we have 33 years.

That's Jesus the Messiah's age when He was crucified, or cut off. So, the last week, or seven years of Daniel's prophecy has not yet been fulfilled.They constitute the seven years of the Great Tribulation Period.

When Jesus died on the cross at the end of the 483 years, this particular prophetical calendar of events was put on hold. There is an indefinite interim period between the 69th week of years and the 70th. This is the period of time reserved by God for building of the Christian Church. No man knows how long exactly this period will last. Only God knows! But this time period will end at the Rapture of the Church and the last week of years will begin with the Great Tribulation Period.

As noted before in a previous chapter, during the first three and one-half years the two witnesses will have their ministry. (Rev. 11:2.) The last three and one-half years will be the appearance and ministry of the Beast, the Antichrist. (Rev. 13:5.) When Daniel's visions were coming to an end, he was told to, "Shut up the words, and seal the book, even to the time of the end: many shall run to and fro, and knowledge shall be increased. Then I Daniel looked, and, behold, there stood other two, the one on this side of the bank of the river, and the other on that side of the bank of the river. And one said to the man clothed in linen, which was upon the waters of the river, How long shall it be to the end of these wonders? I heard the man clothed in linen, which was upon the waters of the river, when he held up his right hand and his left hand unto heaven, and sware by Him that liveth forever that it should be for a time (I year), times (2 years), and a half (6 months); and when he shall have accomplished to scatter the power of the holy people, all these things shall be finished." (Dan.12:4-7.)

All Scripture is given by the inspiration of the Holy Spirit. It is God-breathed. Holy men of old were carried by the Holy Spirit into the very presence of the Sovereign God and wrote and spoke as God possessed every fiber of their beings. And Bible prophecy will happen. It is being fulfilled every day of our lives and we can see the signs of it clearly if we listen to what God has said in the Scriptures and is saying to our hearts. The Tribulation Period will take place and will culminate with the activity of Armageddon.

In the seventh chapter of Daniel when Daniel saw the little horn (authority) rise up from among the Ten Horns (authorities), he described him as "having eyes like the eyes of man, and a mouth speaking great things." (V. 8.) This little horn became a great horn, "whose look was more stout than his fellows." (V. 20.) Let me point you to one more passage of Scripture here. "And he will speak great words against the Most High, and shall wear out the saints of the Most High, and think to change times and laws: and they shall be given into his hand until a time (1 year) and times (2 years) and the dividing of time (6 months). (V. 25.) Folks, this is speaking of the Antichrist and his kingdom. His kingdom will be world-wide. He will rule the world! His speech is blasphemous toward God! His deeds and doings are against God! He is the son of perdition, "Who opposeth and exalteth himself above all that is called God, or that is worshipped; so that he as God sitteth in the temple of God, showing himself that he is God." (II Thes. 2:4.) He is "full of names of blasphemy." (Rev. 17:3)

We need a good definition of the word "blasphemy." Kenneth S. Wuest in his Expanded Translation describes the Beast as being "full of names reproachful and injurious to the divine majesty of Deity." We will see a little later how men reacted to some of the plagues God sent upon them. They "reviled God with impious and reproachful speech injurious to the divine majesty of Deity."

So, Armageddon is the site upon which Jesus will make His appearance when He comes with thousands and thousands of His saints. What are the International events that will make Armageddon necessary?

Paul stated in II Thes. 2:3 that the second coming of the Lord Jesus would not come except there be a falling away first. The falling away to which he referred is a time of apostasy, or an abandonment of Scriptural truth. Actually it will be a mixture of truth and error. In the beginning of the 21st century we have just that. This falling away will continue and intensify until the Rapture of the Church and on into the Great Seven Year Tribulation Period. Indeed, it will be the foundation for the teaching and ministry of the Beast and the False Prophet. The frivolous and flippant church member will be easily fooled into thinking that this sort of teaching is all right, not knowing the true teachings of the Bible.

The Antichrist and False Prophet will have mixed doctrines that side-step the teachings of the Cross and the holiness of God. Paul called it "another gospel." (Gal. 1:6.) This teaching will spill over into the political and social structure of the last society. In fact, it will become mainly a social gospel that has the approval of the lukewarm church and political leaders of the age of the end time.

The International Community, or the New World Order will work so hard to establish some order of peace in the Mid-East and the rest of the world through the establishment of some form of democracy. At the same time they will be uttering impious and reproachful speech and making long-range decisions that are injurious to the divine majesty of Deity, such as making it illegal to pray in public schools or before certain social activities, and condoning homosexual activities. Aren't things like these impious and injurious to the divine majesty of God? There is no lasting peace except that which comes down from above.

Let's face it; it is a religious and social war going on in the Middle East, and Iraq, and Iran, and Korea, and in Liberia. Human rights are emphasized while it is thought to be illegal to emphasize anything Christian. That's an exact description of the ministry and time of the Beast and his False Prophet, except it will be greatly intensified. The nature of the Antichrist is beastly and self-centered. The nature of the False Prophet is the same as the Beast, very self-centered and arrogant and proud. Remember, he will look like a lamb but have the voice and teachings of the Beast.

What does all this mean? The Antichrist will have a world-wide church and it will be an extension of religious mood of the 21st century. Let me refer you to a few Scriptures. "Now the Spirit speaketh expressly, that in the latter times some shall depart from the faith, giving heed to seducing spirits, and doctrines of devils; speaking lies in hypocrisy; having their conscience seared with a hot iron." (I Tim. 4:1,2.) "This know also, that in the last days perilous times shall come. For men shall be lovers of their own selves, covetous, boasters, proud, blasphemers, disobedient to parents, unthankful, unholy, without natural affection, trucebreakers, false accusers, incontinent, fierce, despisers of those that are good, traitors, heady, highminded, lovers of pleasures more than lovers of God; having a form of godlinesss but denying the power thereof: from such turn away. (II Tim. 3:1-5.)

The Antichrist will not only be head of this world-wide church but his political regime will be world-wide, also. He will be The Eighth King and will rule the Eighth World Empire. Isn't that what the Word says? "And the beast that was, and is not, even he is the eighth, and is of the seven, and goeth into perdition." (Rev. 17:11.) The horror of it all is this: that he will be allowed to come working such signs and lying wonders that he will convince the frivolous and ungodly world that he is the Messiah. He will pretend to be setting up the Millennium Kingdom of the Lord Jesus Christ! His incredible deception will be that complete.

The Jewish Nation will believe the Antichrist to be the Messiah and will accept him as such, at least for awhile.The reason for this is because they have never accepted Jesus Christ as the Messiah that should come. They still pray for His coming. I have been to the Wailing or Western Wall and I have seen them praying for Him to come. I have seen them cram their written prayer requests into the cracks and crevices of the wall. I have seen their tears as they pray, some of them kneeling and some standing with uplifted hands, but all praying for the coming of the Jewish Messiah. So, when the Antichrist comes with his plans for peace and his apparently obvious spiritual powers, Israel will accept him as the Messiah, as will the rest of the vast majority of the world's population.

Daniel said, "And he shall confirm the covenant with many for one week: and in the midst of the week he shall cause the sacrifice and the oblation to cease, and for the overspreading of abominations he shall make it desolate, even until the consummation, and that determined shall be poured upon the desolate." (Dan. 9:27.) This Scripture is referring to the Seven Year Tribulation Period. The Jewish Nation is most definitely involved in the International search for world peace. What is the "abomination of desolation" spoken of by Jesus in Matthew 24:15? Jesus referred to it as "spoken of by Daniel the prophet." He also added the words, "whoso readeth, let him understand." This moment in world history is of greatest significance. In this moment God will announce to the Jewish Nation and the world that they have been wrong in their religious calculations. By the coming of this moment in world history the religious and secular world will have such a blend that there will be no distinction between them. The mixture of truth and error will absolutely captivate the mind-set of the race of men.

The Jews rejected Jesus as their Messiah because of His lowly birth and His opposition to their traditions, and for keeping company with sinners. They will accept Antichrist because he comes as a World Leader. He is intelligent and proposes a peace plan that will seemingly work. There is such power and persusion in his presence and person until they will think him to be the Messiah. Everything seems to be going well. World peace seems to be in full swing. There is a spokesman for the regime that seems to have all kind of spiritual powers. (The Bible calls him the False Prophet.) They have seen him call fire down from heaven in the sight of men. He is on top of everything, both religious and secular; that is, until a certain thing happens in Jerusalem. He has engineered a plan to place a statue of the Antichrist in the Temple Square. That is the unacceptable thing in this moment of time in world history. At least, it is unacceptable to the Jew.

The Jewish Nation does believe in the Torah, The Law, and the Law says, "Thou shalt not make unto thee any graven image, or any likeness of anything that is in heaven above, or that is in the earth beneath, or that is in the water under the earth." (Ex. 20:4.) When the False Prophet places the image of the Beast in that sacred spot, and institutes his worship system, Israel is going to rebel against the world system. In fact, she will secede from the International Community. This will not be politically correct. This action will blow the peace process apart. Action must be taken against Israel. So what happens? The International Community with its military forces will surround Israel. They will have one message for her." Back up or we'll blow you away!" Humanly speaking, it doesn't look good for Israel. With the modern technology and the strategies of how to use it, it looks as if this might be the end. Israel will not back up. She has resorted to the Valley of Armageddon, the only battleground she knows. It lies between Jerusalem and the Dead Sea. Israel has fought many battles here before. But this time as world forces are gathered there, it seems that Israel is trapped! But it is at this precise moment in history that something wonderful happens! Let John tell us about it! "And I saw heaven opened, and behold a white horse; and He that sat upon him was called Faithful and True, and in righteousness he does judge and make war. His eyes were as a flame of fire, and on His head were many crowns; and He had a name written, that no man knew, but He Himself. And He was clothed with a vesture dipped in blood: and His name is called The Word of God. And the armies that were in heaven followed Him upon white horses, clothed in fine linen, white and clean. And out of His mouth goeth a sharp sword, that with it He should smite the nations: and he shall rule them with a rod of iron: and He treadeth the winepress of the fierceness and wrath of Almighty God. And He hath on His vesture and on His thigh a name written, KING OF KINGS, AND LORD OF LORDS. (Rev. 19:11-16.)

Our Lord does not fire one shot, neither does His armies that follow Him from heaven. Not one bomb is dropped, nor one missile fired. His holy presence is enough to win the battle. You see, it is God who is descending from heaven to rescue Israel! He told Moses at the burning bush His Name. "I Am that I Am," He had said. When Israel asks for My Name. tell them, "I Am that I Am has sent you! That's My Name! In the Hebrew translation it's a little longer than that. "I shall always be what I always have been!" That's His Name!

Humankind in his earthly form is not able to stand in the holy presence of our Lord. This has always been so. At the giving of the Law on Mount Sinai, the mountain quaked greatly, the voice of the trumpet sounded louder and louder, and the mountain smoked as if it were on fire, and the people were terrified. "And all the people saw the thunderings, and the lightnings, and the noise of the trumpet, and the mountain smoking: and when the people saw it, they removed, and stood afar off. And they said unto Moses, Speak thou with us, and we will hear: but let not God speak with us, lest we die." (Ex. 20: 18, 19.)

In the 10th Chapter of Daniel, he tells us of seeing a vision of the glory of God and his reaction to it. He tells us all strength left him and he had no breath left. He had to be strengthened before he could stand. At the visions of John in Revelation. He tells us of "falling as a dead man." Natural man cannot stand in the holy presence of Deity. It is at Armageddon where strong men's strength will fail them. Mighty military warriors will fall prostrate at His feet. It is here that "every knee will bow and every tongue will confess that Jesus Christ is Lord." God has warned all generations that His Day Is Coming. He spoke through Ezekiel, a prophet and priest of the Babylonian Captivity. "And it shall be when they say unto thee, Wherefore sighest thou? That thou shalt answer, For the tidings, because it cometh: and every heart shall melt, and all hands shall be feeble, and every spirit shall faint, and all knees shall be weak as water: Behold, it cometh, and shall be brought to pass, saith the Lord God." (Ezek. 21: 7.)

At Armageddon the armies of the world will be gathered against Israel, precisely because Israel refuses to do homage to the International Leader, the Antichrist. Israel was cured of idol worship in the Babylonian Captivity. She was scattered to all nations at that time. God has been gathering Israel back to the Holy Land since

May of 1948. She stands as a Nation again in her own land. The Temple Square, where the Temple used to stand, is the most sacred spot on the earth to her. Say what we may about Israel's regard for God, but she does believe in the Law of Moses and The Ten Commandments. It is the placing of the image of the Beast in the Temple Square in Jerusalem that brings about Armageddon. The Battle of Armageddon is set in array by the armies of earthly men. It is Jesus and His armies from heaven that rescues and liberates them. I can hear the Holy Spirit shouting through Paul as he wrote to the Romans, "And so all Israel shall be saved: as it is written, There shall come out of Sion the Deliverer, and shall turn away ungodliness from Jacob: For this is my covenant unto them, when I shall take away their sins." (Rom. 11:26, 27.)

At Armageddon the World System that has left God out of all its plans and strategies will crumble. It will be no more! While it is true the armies of heaven will not fire a shot, or drop a bomb, or shoot a missile, Jesus will use the elements to destroy the System. I will mention two things that will happen as the events of Armageddon take place. There will be the greatest earthquake the world has ever known and a great hail storm. Let the Scriptures speak! "And the seventh angel poured out his vial into the air; and there came a great voice out of the temple of heaven, from the throne, saying, It is done. And there were voices, and thunders, and lightnings, and there was a great earthquake, such as was not since men were upon the earth, so mighty an earthquake , and so great. And the great city was divided into three parts, and the cities of the nations fell: and great Babylon came in remembrance before God, to give unto her the cup of the wine of the fierceness of His wrath. And every island fled away, and the mountains were not found. And there fell upon men a great hail out of heaven, every stone about the weight of a talent: and men blasphemed God because of the plague of the hail; for the plague thereof was exceeding great." (Rev. 16:17-21.)

Let us consider for a moment this great hail storm. You notice, the Bible says every stone was about the weight of a talent. How much does a talent weigh? The Amplified Bible says 50 to 60 pounds; the NIV Bible says about 100 pounds; the New Jerusalem Bible says about 88 pounds each; and J.A. Seiss gives four weights in the Apocalypse: 56, 115, 135, and 390 pounds each, depending on which kind of talent one may be considering. To take the least weight, 50 to 60 pounds, the results would be totally disastrous. You would think that men in such circumstances would repent, wouldn't you? But they did not! Rather they blasphemed God because of the plague of the hail. Kenneth S. Wuest, the Greek scholar, gives an interesting definition of blasphemy. I have referred to this before. "And the men reviled God with impious and reproachful speech injurious to the divine majesty of Deity."

The World System, commonly called The New World Order, or The Global Community is going to collapse and be no more, and a New System shall take its place, the blessed Kingdom of the Lord Jesus Christ. At Armageddon, the Beast and his False Prophet will be cast headlong into hell where they will stay forever. The Great Red Dragon, the devil, will be put into the abyss, bottomless pit, for a thousand years. What glory awaits the believer, and peace! It is a peace that passeth all understanding!

Chapter Nine

The Milennial Reign

Blessed and holy is he that hath part in the first resurrection: on such the second death hath no power, but they shall be priests of God and of Christ, and shall reign with Him a thousand years. (Rev. 20:6. KJV)

After Satan is bound by the mighty angel and cast into the bottomless pit, and the Beast and False Prophet are cast into the lake of fire and brimstone, there will be a thousand years of peace. Jesus Christ will reign from Jerusalem with His saints. In the last chapter we saw Him coming down from heaven to Armageddon with thousands and thousands of His saints. They were riding white horses and arrayed in white robes. They are called the "armies of heaven." They are made up of the righteous of all the generations of the earth, from Abel to Armageddon. Abel is in that blessed group; so is Abraham, Isaac and Jacob, and David, and the prophets, and the New Testament saints. Peter and Paul; James and John, and all others who have made their robes white in the blood of the Lamb. The Bible says that we will reign with Him a thousand years. What will that reign be like?

First of all, the government of Jesus Christ will be

centered in Jeursulem, and He will rule the Nations of the world from that location. "And the government shall be upon His shoulder: and His name shall be called Wonderful, Counsellor, The Mighty God, The Everlasting Father, The Prince of Peace. Of the increase of His government and peace there shall be no end, upon the throne of David and upon his kingdom, to order it, and to publish it with judgment and justice from henceforth even forever. The zeal of the Lord of hosts will perform this." (Isa. 9:6, 7.) What people will comprise the population of these nations? From whence do they come? They are the remnant who survived the Great Tribulation Period which ended at Armageddon. They are Christians who survived the reign of the Beast. God has reserved them to re-plenish the earth in the thousand year Millenium. They will fill the earth with a generation of people who are free from the temptation of the devil.. Satan and the Beast wanted to destroy them during the reign of the Antichrist but could not because God had prepared a place for them in the wilderness or desert, and He protected them there. "And when the dragon saw that he was cast unto the earth, he persecuted the woman which brought forth the man child. And to the woman were given two wings of a great eagle, that she might fly into the wilderness, into her place where she is nourished for a time (one year), and times (two years), and half a time (six months), from the face of the serpent. (This three and one half year period is the length of the Great Tribulation Period). And the serpent cast out of his mouth water as a flood after the woman, that he might cause her to be carried away of the flood. And the earth helped the woman, and the earth opened her mouth, and swallowed up the flood which the dragon cast out of his mouth. And the dragon was wroth with the woman, and went to make war with the remnant of her seed, which keep the commandments of God, and have the testimony of Jesus Christ." (Rev. 12:13-17.) This remnant will be men and women of flesh, bone and blood. They have never been raptured and they shall be the parents of the generations born during the Millennium Reign.

We must keep in mind that we are discussing the Millenial Reign, not the events that will occur after the Consummation. We will discuss the Consummation in a later Chapter. There is a great distinction.

There are special promises made in the Bible to the overcomers of the Great Tribulation Period, and there are special phrases used that describe the times of the Millennial Reign of Christ. Jesus said of the overcomer in the Thyatiran Church Age, "And he that overcometh and keepeth my works unto the end, to him will I give power over the nations: and he shall rule them with a rod of iron." (Rev. 2:26,27.) We are told in Rev. 12:5 of the woman that brought forth the man- child, "And she brought forth a man child, who was to rule all nations with a rod of iron." Of course, as we look at this, we must recognize that the woman is the Visible Organized Church and her offspring is the Invisible Church which is nestled within and brought forth by the Visible Organized Church. The Invisible Church is the True Church, the Body of Christ, the Church Triumphant, which is mentioned so often in the Scriptures. Who are those that shall rule and reign with Christ? They are the raptured saints who return with Him to Armageddon. They shall set up the Kingdom and rule the Nations of the world from Jerusalem.

There are some things that we must consider concerning the Millennial Kingdom. For one thing, it is the Reserved Remnant that re-populate the earth after Armageddon, for they are men and women of flesh, bone,and blood. It is the Raptured Saints that shall rule from Jerusalem. They have Raptured bodies of flesh, bone and Spirit. They have undergone the glorious change that occurs at the Rapture!

The chief characteristic of the Millennial Reign will be Holiness Unto The Lord. Perfection shall reign. Pure justice and true judgment will always be in view. But there will still be Nations on the earth. They shall bring their glory and riches and beauty into Jerusalem. If there are Nations, there will be heads of Nations with their governmental bodies. They will serve in the International Community of that time in a vassal capacity. If any man or Nation steps out of line, they will be brought back into line by the perfect Reign of Jesus Christ and His saints. We can be assured that God will not take away man's ability to choose, his Free Moral Agency, after Armageddon.There will be no wars or rumors of wars because of His blessed presence on earth. "The word that Isaiah the son of Amoz saw concerning Judah and Jerusalem. And it shall come to pass in the last days, that the mountain of the Lord's house shall be established in the top of the mountains, and shall be exalted above the hills; and all nations shall flow into it. And many people shall go and say, Come ye, and let us go up to the mountain of the Lord, to the house of the God of Jacob; and He will teach us of His ways, and we will walk in His paths: for out of Zion shall go forth the law, and the word of the Lord from Jerusalem. And He shall judge among the nations, and shall rebuke many people: and they shall beat their swords into plowshares, and their spears into pruninghooks: nation shall not lift up sword against nation, neither shall they learn war anymore. O house of Jacob, come ye, and let us walk in the light of the Lord." (Isa. 2:1-5.) This pretty much describes the environment the people of the Millenial Reign will find themselves in.

Many things will be changed after Armageddon, but the order of Nations and their governments will continue. I must emphasize, though, that Jesus, our Lord will be Head of all. His Kingdom will be world-wide and a Kingdom of Peace. The watchword and character of His Kingdom will be Holiness Unto The Lord! "In that day shall there be upon the bells of the horses, HOLINESS UNTO THE LORD; and the pots of the Lord's house shall be like the bowls before the altar. Yea, every pot in Jerusalem and in Judah shall be holiness unto the Lord of hosts." (Zech. 14: 20, 21.)

As the world, the whole International Community, enjoys the blessings and grace of this Millennial Kingdom of Christ, the length of life will be extended just as it was before the flood. The generations of Adam are listed in the 5th Chapter of Genesis. Men lived in those days hundreds of years. Methuselah, the grandfather of Noah, lived the longest of any listed in that genealogy. It is stated that Methuselah lived to be 969 years old. Noah lived to be 950 years old.

God cut the longevity of man's years back just before the flood when the earth became filled with violence by mankind's own doings. The story is recorded in the 6th Chapter of Genesis. This is a portion of Scripture that is greatly misunderstood by some Bible commentators. The Scripture says, "And it came to pass, when men began to multiply on the face of the earth, and daughters were born unto them, That the sons of God (descendants of Seth) saw that the daughters of men (descendants of Cain) that they were fair; and they took them wives of all which they chose." (Gen. 6:1, 2.) The view that these sons of God were angels is incorrect. Angels are Spiritual, heavenly Beings, and they have no interest in marriage at all, or intimately intermingling with earthly females of flesh and blood. You see, Seth represents the Godly lineage of the human race while Cain represents the ungodly lineage of the human race. God did not interfere with their Free Moral Agency in their choices but He did intervene in the results of those choices. He saw the detriment and the direction to which their choices were leading them. And He simply stopped them! "And God saw that the wickedness of man was great in the earth, and that every imagination of the thoughts of his heart were only evil continually...," He made a statement. "And the Lord said, My Spirit shall not always strive with man, for that he also is flesh: yet his days will be a hundred and twenty years." (Gen.6:3.) So God shortened man's life span, appointed Noah to build the ark, and brought the flood on the earth.

Genesis 6:10 tells us, "And Noah begat three sons, Shem, Ham, and Japheth." Of these three Shem was representative of the Godly lineage and Ham became the representative of the ungodly lineage. From Ham came Cush, and from Cush, Nimrod, who was the progenitor of the World System that has always rebelled against God. Of Nimrod's generation the Bible says, "And the Lord came down to see the city and the tower, which the children of men builded. And the Lord said, behold, the people is one, and they have all one language; and this they begin to do: and now nothing shall be restrained from them, which they have imagined to do. Go to, let us go down, and there confound their language, that they may not understand one another's speech. So the Lord scattered them abroad from thence upon the face of all the earth and they left off to build the city." (Gen. 11:5-8.)

God once again shortened man's longevity. Psalm 90:10 tells us, "The days of our years are threescore years and ten; and if by reasons of strength they be fourscore years, yet is their strength labor and sorrow; for it is soon cut off and we fly away." But in the Millennial Reign of Jesus, the Christ, our longevity will be restored as it was before the flood. The Lord says, "And I will rejoice in Jerusalem, and will be glad in my people: and there will no more be heard in her the voice of weeping, or the voice of crying. Neither shall there be there any more a child that dies untimely, nor an old man who shall not complete his time: for the youth shall be a hundred years old, and the sinner who dies at a hundred years shall also be accursed: and they shall build houses, and themselves shall dwell in them; and they shall plant vineyards, and themselves shall eat the fruit thereof." (Isa. 65:19-21. The Septuagint Version: Greek and English.) This is a Scriptural discussion of the times of the Millennial Reign, not the Consummation. We must notice that the "sinner" and "dying" is mentioned by the prophets as occurring in that Thousand Years of

Peace. It will be a glorious time of perfect peace. Jesus Christ will be seen in Jerusalem and His raptured saints will be in charge of His perfect government. A Utopia? Yes, but as real as heaven itself.

God promised through the prophet Hosea, a contemporary of Isaiah, "And in that day will I make a covenant for them with the beasts of the field..." (Hos. 2:18a.) God also describes through Isaiah, His prophet, how the covenant works in the times of the Millennial Reign of Christ. "The wolf also shall dwell with the lamb, and the leopard shall lie down with the kid; and the calf and the young lion and the fatling together; and a little child shall lead them. And the cow and the bear shall feed; their young ones shall lie down together: and the lion shall eat straw like the ox. And the sucking child shall play on the hole of the asp, and the weaned child shall put his hand on the cockatrice' den. They shall not hurt nor destroy in all my holy mountain (kingdom): for the earth shall be full of the knowledge of the Lord, as the waters cover the sea." (Isa. 11:6-9.)

What a blessed time period the Millennial Reign of our Lord Jesus Christ will be! There will be a thousand years of peace on earth. In one sense, it will be a thousand year Sabbath, if we take literally what the Apostle Peter said. "Knowing this first, that there shall come in the last days scoffers, walking after their own lusts, And saying, Where is the promise of His coming? For since the fathers fell asleep, all things continue as they were from the beginning of creation. For this they willingly are ignorant of, that by the word of God the heavens were of old, and the earth standing out of the water and in the water: Whereby the world that then was, being overflowed by water, perished: But the heavens and the earth which are now, by the same word are kept in store, reserved unto fire against the day of judgment and the perdition of ungodly men. But, beloved, be not ignorant of this one thing, that one day is with the Lord as a thousand years, and a thousand years as one day. The Lord is not slack concerning His promises, as some men count slackness; but is longsuffering to us-ward, not willing that any should perish, but that all should come to repentance." (II Peter 3:3-9.)

Let's look at a week of a thousand years per day. From creation to the flood was two thousand years. Let's place these days into the Christian Calendar. The first two thousand years is Monday and Tuesday. From the flood to the first Advent of Jesus was two thousand more years. That will be Wednesday and Thursday of our week, Four thousand years have passed. From the first Advent of Jesus Christ to Armageddon is two thousand more years. That will be Friday and Saturday of our week. That's six days of our thousand years per day week. The Millennial Reign is the seventh thousand year day, giving it the character of a thousand year Sabbath Day. This theory is interesting, isn't it? Let me say this. Time with God is not what it is with man. This is for certain!

The thousand year Millennium as an appointed time period with God must come to an end, and it does with a jolt. And then the Judgment!

Chapter Ten

The Battle of Gog and Magog

And when the thousand years are expired, Satan shall be loosed out of his prison, And shall go out to deceive the nations which are in the four quarters of the earth, Gog and Magog, to gather them together to battle: the number of them is as the sand of the sea. And they went up on the breadth of the earth, and compassed the camp of the saints about, and the beloved city: and fire came down from God out of heaven, and devoured them. (Rev. 20:7-9. KJV)

You'll notice at the first of the above text that the Bible says the thousand years expired. The time of the thousand year Sabbath is ended. It is now time for another segment of time to begin which will last only a short time.

We must remember that God is always knowledgeable of events in heaven and events on the earth. Nothing happens without His awareness. God's knowledge is absolute concerning the past, present, and future. In fact, He is absolute in all of His divine attributes. He is Omnipotent, meaning He is Almighty. He is Omniscient, meaning He knows everything all the time, past, present and future, He is Omnipresent, meaning He is everywhere at all times. There is nothing too hard for Him to do. He is full of grace and truth, He is a Holy God.

The beauty of it all is that He created man in His own image, after His own likeness. This does not mean that God is a man, nor man a god, other than in the Person of the Lord Jesus Christ. God is a Spirit, according to what the Lord Jesus told the Samaritan woman at Jacob's well. God made man in His own image in that he gave him the ability to reason and make good judgment, Also, God made man an eternal creature, to live forever. He made man an intelligent creature, the most intelligent of all His creation. He made man with a Free Moral Agency, the power and ability to choose, to make choices between right and wrong. God did not make man as a robot, compelled to love Him and serve Him. He wanted man to choose of his own volition to love and serve Him. Perhaps the best way to say it is that God made man with the right to choose and He will not infringe upon that right. Regardless of man's choices, good or bad, God will not take his power to choose away from him. He will eventually judge man for his choices and works. God made the earth and all that's in it and gave it to man to govern. Whatever state it may be in at any given time will be largely due to the result of man's decisions and choices.

We have already seen that it will be the doings of man in his International Community that bring about Armageddon. We must now turn our attention to another battle. That is, the battle of Gog and Magog. This is the final battle of time as we know it. It takes place at the end of the Millennial Reign of our Lord Jesus Christ.

The question might be asked, "Why is this battle necessary?" There has been a thousand years of peace on the earth. Why Gog And Magog? The answer is quite simple. Satan has been bound in the bottomless pit during the Thousand Year Millennial Reign and has not, therefore, been free to tempt mankind with his evil devices. Until Armageddon, he was loose in the realm of the earth to put forth his temptations and evil schemes. All men from Eden until Armageddon had been his potential prey. Some had resisted him and some had fallen victim to him. During the Thousand Year Millennial many people will be born into the world. Their numbers are as innumerable as the sands of the sea shore. Until the release of Satan from his prison, they will not have undergone his onslaught of lies, accusations, and evil character. God is just and impartial! He cannot in His divine justice allow all other generations to undergo the temptation of Satan's wiles, and allow the Millennial generations to be free of them. This would not be just. This is the reason for Satan's release, to give the Millenial generations a choice. This was the reason for the temptation in the Garden of Eden. You see, man must make his own decision to follow the Master and do justly.

There is no good in Satan. After his release in post millennial days, he will take up where he left off. He will move upon the International Community with all speed and through his influence upon World leaders will amass an innumerable army, made up of men of every nation on the earth. His effort will most certainly be on the International level and his goal will be, as it always has been, and that is to destroy the Kingdom of God.

The Battle of Gog and Magog will be Satan's last

attempt to overthrow the Kingdom of God. It will be his last revolt against God's holy plan to redeem all the human race from sin and its consequences. Satan does not know everything. Some things are still a mystery to him. But one thing he does know, and that is , that his time is short. He knows that hell awaits him. He knows that hell was prepared, not for man, but for him and his angels. All of the inhabitants of hell know this. When Christ ministered to man in His own earthly ministry, devils cried out, "And, behold, they cried out, saying, What have we to do with thee, Jesus, thou Son of God? Art thou come hither to torment us before the time?" (Matthew 8:29.) "And there was in their synagogue a man with an unclean spirit; and he cried out, Saying, let us alone; what have we to do with thee, thou Jesus of Nazareth? Art thou come to destroy us? I know thee who thou art, the Holy One of God." (Mark 1:23, 24.) So, at Satan's release from the abyss, the bottomless pit, he knows that his time is very short. Therefore, he will do his worst. The New Jerusalem Bible says that it is his last revolt against God and His Kingdom.

Since Satan's fall, he has been the same, a rebel against God, Jesus said, "I beheld Satan as lightning fall from heaven." (Luke 10:18.) He also said, "He (Satan) was a murderer from the beginning. and abode not in the truth, because there is no truth in him. When he speaketh a lie, he speaketh of his own: for he is a liar. and the father of it." (John 8:44.) Satan has not changed. Even though he knows his time is short, he will stir up all the mischief and evil he can. Therefore, when he is released from the bottomless pit, he will organize a massive, International Army whose number cannot be counted for multitude. The purpose is to overthrow the earthly Kingdom of Jesus Christ and destroy the Kingdom of God. Daniel saw this attempt happening in his visions and placed it just before the resurrection of the wicked. (See Dan. 12: 1,2.) Joel saw the International Army, amassed and in action. Let him tell it as he saw it. "Blow ye the trumpet in Zion, and sound an alarm in my holy mountain: let all inhabitants of the land tremble: for the day of the Lord cometh, for it is nigh at hand; And a day of darkness and of gloominess, a day of clouds and of thick darkness, as the morning spread upon the mountains: a great people and a strong; there hath not been ever the like, neither shall be anymore after it, even to the years of many gernerations. A fire devoureth before them; and behind them a flame burneth: the land is as the Garden of Eden before them, and behind them a desolate wilderness; yea, and nothing shall escape them. The appearance of them is as the appearance of horses; and as horsemen, so shall they run. Like the noise of chariots on the tops of mountains, shall they leap, like the noise of a flame of fire that devoureth the stubble , as a strong people set in battle array. Before their face the people shall be much pained: all faces shall gather blackness. They shall run like mighty men: they shall climb the wall like men of war; and they shall march every one on his ways, and they shall not break their ranks: Neither shall one thrust another; they shall walk every one in his path: and when they shall fall upon the sword, they shall not be wounded. They shall run to and fro in the city; they shall run upon the wall, they shall climb up upon the houses; they shall enter in at the windows like a thief. The earth shall quake before them; the heavens shall tremble: the sun and the moon shall be dark, and the stars shall withdraw their shining." (Joel 2:1-10.)

This massive International Army shall surround Jerusalem. This is not the first time Jerusalem has been surrounded by the International Military Forces of nations. Just a thousand years ago it had happened at Armageddon. Satan was its instigator then and so he is now. Again, let me remind you, it is men or mankind that bring these woes upon themselves by the choices they make. But in every instance and circumstance God is calling out to whosoever will hear Him for the better way of repentance and forgiveness. This massive International Force will destroy the beauty of Eden generated by the Millennial Reign of our Lord Jesus Christ. The Scripture said that before them the appearance was as the garden of Eden, but behind them was burned and charred desolation. I'm sure you noticed the supernatural character and carriage of this devastating army.

But the All-Wise, Sovereign God is looking on and He sees the tremendous loss the choices of men have brought upon themselves and He intervenes once again. Revelation, Chapter 20, Verse 9 says,"And they went up on the breadth of the earth, and compassed the camp of the saints about, and the beloved city: and fire came down from God out of heaven, and devoured them."

Listen, as the prophet Zechariah puts it another way, "And this shall be the plague wherewith the Lord will smite all people who have fought against Jerusalem; their flesh shall consume away while they stand upon their feet, and their eyes shall consume away in their holes, and their tongue shall consume away in their mouth." (Zech. 14:12.)

But there is a new day coming! In our writings, the ultimate end is not too far away. But there are other things to happen before that can take place. Dear reader, let me remind you! Death at the end of this earthly life is not the end! It is really the beginning of a broader, more beautiful life. God has made a lot of promises to you and He reserves eternal life for you and all its glory; that is, if you will only make the right choice and enter into His fellowship right now!

At the end of Battle of Gog and Magog, God will deal with the devil once and for all. He will cast him into the lake of fire and brimstone, where he will remain forever. The Beast and False Prophet are already there. They have been there a thousand years for they were cast there at the battle of Armageddon. Read about the doom of Satan in Rev. 20, Verse 10.

There is one last thing we must consider before we get to the ultimate end and we will consider that in the next Chapter.

Chapter Eleven

The Judgment

And I saw a great white throne, and him that sat on it, from whose face the earth and the heaven fled away; and there was found no place for them. And I saw the dead, both small and great, stand before God; and the books were opened: and another book was opened, which is the book of life: and the dead were judged out of those things which were written in the books, according to their works. And the sea gave up the dead which were in it; and death and hell delivered up the dead which were in them: and they were judged every man according to their works. And death and hell were cast into the lake of fire. This is the second death. And whosoever was not found written in the book of life was cast into the lake of fire. (Rev. 20:11-15.)

When God formed Adam's body, we can be sure it was the perfect specimen of the male species of the human race, for God's creation in the beginning was flawless. Adam became a living soul when God breathed the breath of life into him. That's when the body in all its entirety began its functions. I remind you once again that the tangible, physical body has three main components, which are flesh, bone and blood. These components did not begin to function until God breathed the breath of life into the man, Adam. The eternal being, Adam, began living at that moment.

Adam was not only comprised of flesh, bone and blood, the tangible physical creature, but he was also an intangible, spiritual creature comprised of soul, mind and spirit. This is the real Adam who is living in the physical body. Peter referred to this body as a tabernacle, a mere tent, in which he temporarily lived. "Yea, I think it meet, as long as I am in this tabernacle, to stir you up by putting you in remembrance; Knowing that shortly I must put off this tabernacle, even as the Lord Jesus Christ hath showed me." (II Peter 1:13,14.) So, Adam, the inner man will live forever! And all Adam's descendants will live forever! This is one facet of the image in which God made him "after His likeness." As God is eternal, so is His creature, Adam.

If we believe God made Adam physically perfect, then we must believe He made him spiritually perfect, also. We do know that Adam and Eve had very close fellowship with God, As they walked through the Garden, they had verbal communion with Him. The Bible says, "And they heard the voice of the Lord God walking in the garden in the cool of the day: and Adam and his wife hid themselves from the presence of the Lord God amongst the trees of the Garden." (Gen. 3:8.)

Adam began his life in the Garden as a whole man, both physically and spiritual. We can only imagine the completeness of his intellect at first. Whatever degree it might have been, when he disobeyed God there was a great shadow that greatly dimmed its brightness. His great sin was bound in not doing the revealed will of the Lord God, but his own will. At that moment selfishness replaced godliness, impurity replaced purity, and personal sin replaced holiness. God had made him capable of ruling the world, but at his fall, division and personal blame filled his household. Adam had become a sinner! His personal nature was changed. There was now a nature within him demanding his own selfish will be done, and not God's will. This fallen nature was greater than Adam. He had alienated himself from his God.

In today's society, psychologists and theologians are busy developing what they call a western theology. They have spawned the idea that the Western Civilization needs a new theology, beginning not with God but with man. They refer to the biblical account of the Garden of Eden merely as "the myth of the garden." And they have termed Satan as "self" or "the self-will of man." This new theology teaches that mankind is "estranged" from God, not lost. They want to refer to God as the "essential being" and man as "existential being."

There is an incident recorded in the Gospels concerning a hypothetical story the Saducees told Jesus about a Jew who married a woman and died leaving no son to carry on his name. This Jewish man had six brothers, so the second brother married the woman to raise up seed to his deceased brother. He died also, leaving no son. This went on until all the seven brothers had her to wife. Their question? In the resurrection, whose wife shall she be, seeing they all had her. Let's hear Jesus' answer, "Do ye not therefore err, because ye know not the Scriptures, neither the power of God? For when they shall rise from the dead, they neither marry or are given in marriage, but are as the angels which are in heaven." (Mark 12:24,25.) Based on the answer of Jesus, the new theology determines that since there are no sexes or genders in heaven, then we don't know if God is male, female, or neuter gender. Therefore, they refer to Him as He, or She, or It. Let me say one thing here. Jesus referred to Him as His Father, and so will I!

It was in the Garden scene that the first mention of a Redeemer is made. God told the serpent, "And I will put enmity between thee and the woman, and between thy seed and her seed; it shall bruise thy head, and thou shalt bruise his heel." (Gen. 3:15.) In this passage of Scripture, Jesus Christ is promised as the Redeemer of Adam's fallen race. He is the Seed of the woman, not the man. The Immaculate Conception is referred to here. If the Bible account of the Garden of Eden is a myth, then this promise is also a myth. This cannot be! It is in these foundational Scriptures that the whole plan of man's redemption is laid!

So, the first man, Adam, sinned through his disobedience to the revealed will of God and brought universal sin into the world. Sin has a penalty. It must be judged, then comes the penalty. The penalty has always been death. Isn't that what God told Adam? "But of the tree of the knowledge of good and evil, thou shalt not eat of it: for in the day that thou eatest thereof thou shalt surely die." (Gen. 2:17.) This is not referring to physical death, but separation from the intimate fellowship with God. With the absence of that fellowship comes many of the errors and disappointments of life.

In the Old Testament, God accepted the death of the sacrificial animals and fowls as being sufficient for the penalty, the innocent dying for the guilty. Each sacrifice was a type of the Ultimate Sacrifice, Jesus Christ. At the Cross of Calvary this Ultimate Sacrifice was made. In the ultimate sense it was the Innocent dying for the guilty.

The pentinent sinner is judged for all his sins at the Cross. Not only that, but he is acquitted of all guilt, even the inherited Adamic nature. He is declared guilt-free! The old Adamic nature is taken away and he is given the Christ-nature. He is born again! The very source and seat of his desires is changed. He desires only to please God. The Fellowship is restored. The love and joy of the Lord God again flows through his being and he is a changed creature.

But Jesus said, "Many will be called but few chosen." Some will shun the call of God to repentance and change. Man fell from his lofty estate in the Garden. A great shadow engulfed him and he cannot see clearly anymore. I am not talking of physical vision. I am talking of the spiritual vision. Even at his best, he is "seeing through a glass darkly." But any man can kneel at the cross, confess his sins and walk away justified , knowing his sins are forgiven. God throws his sins behind His back, into His sea of forgetfulness, never to remember them against him any more. That man has been judged and stands justified and acquitted before God.

Let's look at God's love for a moment. The Bible says, "The Lord hath appeared of old unto me, saying, Yea, I have loved thee with an everlasting love: therefore with lovingkindness have I drawn thee." (Jer. 31:3.) We might ask the question, "When did God begin to love me?" The answer is startling! God never "began" to love me, because He always has. When God appeared to Moses in the burning bush and gave him orders to go to Egypt and liberate the people of Israel, Moses asked Him a question. "Behold, when I come unto the children of Israel, and shall say unto them. The God of your fathers hath sent me unto you; and they shall say to me, What is His name? What will I say unto them? And God said unto Moses, I AM THAT I AM: and He said, Thus shalt thou say unto the children of Israel, I AM hath sent me unto you." (Ex. 3:13,14.) I found what that Name really means as I read in the Torah. "I shall always be what I always have been." That's His Name! God has no beginning and He has no ending. He is eternal. And so are all His divine attributes. More will be said about this later.

There is another illustration of God's judgment I want to mention. During the forty years of wilderness wanderings of Israel, there was a time when they went up to Edom by way of the Red Sea. God had just given them a great victory over the Caananites, but they became greatly discouraged and depressed because of the trials of the way and began to complain and murmur against God and against Moses. They said, "Why have you brought us out of Egypt to die in the wilderness? For there is no bread, neither is there any water, and we loathe this light (contemptible, unsubstantial) manna. And the Lord sent fiery (burning) serpents among the people, and they bit the people, and many Israelites died. And the people came to Moses, and said, We have sinned because we have spoken against the Lord and against you; pray to the Lord, that He may take away the serpents from us. So Moses prayed for the people.

And the Lord said to Moses, Make a fiery serpent [of bronze], and set it on a pole,; and every one that is bitten, when he looks on it, shall live. And Moses made a serpent of bronze and put it on a pole, and if a serpent had bitten any man, when he looked to the serpent of bronze [attentively, expectantly, with a steady and absorbing gaze], he lived." (Numbers 21:5-9. Amplified Bible.)

We must understand fully what the Bible is saying in the above passage. "Jesus said that as Moses lifted up the serpent in the wilderness, so must the Son of Man be lifted up, "that every one who believes in Him---who cleaves to Him, trusts and relies on Him---may not perish, but have eternal life and [actually] live for ever!" (John 3:14,15.) Obviously this implies that the look that caused the victim of a fiery serpent to be healed was something far more than a casual glance. A "look" would save, but what kind of look? The Hebrew text here means "look attentatively, expectantly, with a steady and absorbing gaze." Or as Jesus said in the last verse of the chapter quoted above, "He that believes on --- has faith in, clings to, relies on --- the Son has (now possesses) eternal life." But whoever does not so believe in, cling to, rely on the Son, will never see... life." The look that saves is not a fleeting glance; it is a God-honoring , God-answered, fixed and absorbing gaze! (Footnote on Numbers 21:9. Amplified Bible.)

Adam's race has, indeed, been bitten by the the serpent and is afflicted by the poisonous venom of sin. This is the sin nature that mankind took upon himself at the fall in the Garden of Eden. All works of sin spring from it. Jesus came to destroy the works of sin. It is at the

Cross where all sin and this sin nature is judged. If any man will steadfastly consider his sinful state, fully realizing he is a sinner by nature, he can come to the Cross and behold the Son of God hanging there, and know it is for him, the sinner, that He has been put there by God's Divine mercy, he can be forgiven for his sins and healed of his sin nature. He is saved at the Cross. He has a new nature! And he lives!

Jesus very plainly said, "He that believeth on Him is not condemned: but he that believeth not is condemned already, because he hath not believed in the name of the only begotten Son of God." (John 3:18,) If we apply the same Hebrew meaning of "believing on Him" to the unbeliever, we must know that he that does not believe on Him has never relied upon Him by "attentatively, expectantly, with a steady and absorbing gaze." He has, therefore, shunned the healing of the soul. Let me state here that there is a very, very critical experience of salvation that takes place when a man repents before God. He, indeed, is a changed creature and all things become new. This newness of life comes, not by anything he has done, but by the absolute and powerful work of grace.

We must now come to the final judgment of those who have rejected the call of God to repentance. It is spoken of everywhere in the Bible, in both the Old Testament and the New. Daniel described it like this: "I beheld till the thrones were cast down, and the Ancient of Days did sit, whose garment was white as snow, and the hair of His head like the pure wool: His throne was like the fiery flame, and His wheels as burning fire. A fiery stream issued and came forth from before Him:

thousand thousands ministered unto Him, and ten thousand times ten thousand stood before Him: the judgment was set, and the books were opened." (Dan. 7:9,10.) John saw it much the same way, yet in greater detail. "And I saw a great white throne, and Him that sat on it, from whose face the earth and the heaven fled away; and there was found no place for them. And I saw the dead, small and great, stand before God; and the books were opened: and another book was opened, which is the book of life: and the dead were judged out of those things which were written in the books, according to their works. And the sea gave up the dead which were in it; and death and hell delivered up the dead which were in them: and they were judged every man according to their works. And death and hell were cast into the lake of fire. This is the second death. And whosoever was not found written in the book of life was cast into the lake of fire." (Rev. 20:11-15.)

Time, as we know it, is almost up, as depicted in these Scriptures. The judgment is set! We must look at those who are called to it to give an account of the things they did while they lived on earth. In reality, every man will be judged on his analysis of the Cross of Calvary. Jesus said in John 3:18, "He that believeth on Him is not condemned, but he that believeth not is condemned already." So, as far as sin is concerned, the sin question is settled at the Cross.

The Great White Throne is the place where the Court of the Sovereign God of all heaven and earth shall be held. What a solemn assembly it will be!The summons will go out to all the wicked dead to assemble, from Adam until that great day. It is a segment of time that's reserved for that particular happening. All those who refused the call of God will be there. They have been held in reserve for this momentous event from their deaths until they receive their summons. The summons will not come to be accepted or rejected. There is no choice offered. They must come before God to be judged according to their works done in the flesh, while they lived on earth. God is just! For centuries He has told the world that "The wages of sin is death, but the gift of God is eternal life through Jesus Christ our Lord." (Rom. 6:23.) The time has come to collect their wages.

There are two resurrections. This is the second. The first general resurrection took place at the Rapture of the Church. There are a thousand years between the Resurrection of the Church and the resurrection of the wicked. The Bible says, "Blessed and holy is he who has part in the first resurrection: on such the second death hath no power, but they shall be priests of God and of Christ, and shall reign with Him a thousand years." (Rev. 20:6.)

You'll notice that the above Scripture mentions "the second death." What is the second death? We will explain this in a little while. Right now, let's behold this great, innumerable company of people standing before God. The great and the small, the king and the pauper stand side by side. The rich and the poor stand shoulder to shoulder. The books are brought out. These books contain the record of each one, regardless of his or her position in life. There is another Book brought out. It is the Book of Life. Why will the Book of Life be there at the judgment of the sinner? Again I say, God is just! The "books" record every deed and act done in the flesh, especially every invitation and call the Lord has given, and the rejection by the sinner. Not one will be left out!The Book of Life is there to be searched in the presence of the accused to convince him that his name is missing. Perhaps there will be a blank spot, where his name was once recorded before it was blotted out. What an awful time it will be for the lost! Some man might fancy to say in his heart, "Well, I'll tell the Lord and He'll understand." But there will be no debate! Remember, judgment is taking place based on those things written in the books. The sinner will be speechless. (Matt. 22:12.) It is an awesome time! There are no chances for salvation beyond this present life.

God is the Judge! His great question will be, "What did you do with My Son? What did you think of Him? I sent Him to earth so that you could be saved from a sinful, self-centered life. He told you He was My Son. Did you believe Him? Or did you think He was lying?

At the Judgment the sinner will remember in detail every time God called him. He'll remember how hard it was to resist the call. He will remember his reasons for saying no to the wooings of the Holy Ghost. Even though he is speechless, he will know that the Judgment of the Holy God is just.

The Book of Life will be searched for the sinner's name. When his name cannot be found in the Book of Life, he will be cast into the lake of fire where he will stay eternally. All hope is gone! There will be no more invitations to an altar of prayer, no more invitational hymns, and no more of the sweet wooings of the Holy Spirit. Someone might say, "God would not cast me into such an awful place! He loves me too much." That is true! He does love the sinner. But it is not God Who casts the sinner into the lake of fire. It is the way the sinner has chosen. He has shunned the Cross and chosen to live in the pleasures of his own self-will. The lake of fire is the ultimate-end of the way he has chosen. The Bible calls it the "second death."

The first death occurs when the soul, mind, and spirit leaves the physical body of man in natural death. The Bible says, "It is appointed unto man once to die, and after death, the judgment." The "second death" occurs at the Great White Throne Judgment bar of God. It is eternal separation from God. There are no appeals and no reprieves! The door to eternal life in heaven is shut and no man can open it.

Dear reader, won't you hear what God is saying and draw very close to Him just now Oh, how He loves you and longs for you to enjoy His great salvation in its complete and greatest extent.

Chapter Twelve

The Consummation and Beyond

And I saw a new heaven and a new earth: for the first heaven and the first earth were passed away; and there was no more sea. And I John saw the holy city, new Jerusalem, coming down from God out of heaven, prepared as a bride adorned for her husband. And I heard a great voice out of heaven saying, Behold, the tabernacle of God is with men, and He will dwell with them, and they shall be His people, and God Himself shall be with them, and be their God. And God shall wipe away all tears from their eyes; and there shall be no more death, neither sorrow, nor crying, neither shall there be any more pain: for the former things are passed away. And He that sat upon the throne said, Behold, I make all things new. And He said unto me, Write: for these words are true and faithful. And He said unto me, It is done. I am Alpha and Omega, the beginning and the end. I will give unto him that is athirst of the fountain of the water of life freely. He that overcometh shall inherit all things; and I will be his God, and he shall be My son. (Rev. 21:1-7.)

The judgment is now past and the Consummation is taking place. God's appointed time to "make all things new" has arrived. The definition for the word "consummation" is, "the ultimate end." Of course, as we will apply this definition in this chapter, it will be the ultimate end of all things as we know them now, but does that mean there will be nothing beyond the ultimate end? Of course not! The Bible gives us enough light and understanding, if we look for it, to have a pretty good idea of what is beyond this life for us. However, I will attempt to distinguish the difference between Bible facts and our own speculations.

First of all, the Bible says, "One generation passeth away, and another generation cometh: but the earth abideth forever." (Eccl. 1:4.) After the flood, God's command to Noah and his sons was to be fruitful and multiply and replenish the earth, More specifically, "God said, This is the token of the covenant that I make between Me and you and every living creature that is with you, for perpetual generations: I do set My bow in the cloud, and it shall be for a token of a covenant between Me and the earth." (Gen. 9:12,13.) "Perpetual generations" here means everlasting generations.

We must again go back to the Garden of Eden for a moment and hear the words of the Lord God as He talked with Adam and Eve. "So God created man in His own image, in the image of God created He him; male and female created He them. And God blessed them, and said unto them, Be fruitful, and multiply, and replenish the earth, and subdue it: and have dominion over all the fish of the sea, and over the fowl of the air, and over every thing that creepeth on the earth." (Gen. 1:27,28.) Before Adam could begin the process of filling the earth with his offspring, he sinned through disobedience and fell. As we know, he eventually did fill the earth, but with a fallen race. You see, the command of God to fill the earth with his offspring was while Adam was yet pure and perfect and innocent. His dominion over all the earth and its creatures was given while he was in this perfection and innocency. To paraphrase a little, God was telling Adam to extend Eden and its beauty all over the face of the earth.

Satan's visit to Adam and Eve in the Garden did not change God's charge to Adam, but it did change Adam's nature from holiness and innocency to sinfulness and rebellion. Adam became a fallen creature and his world around him became fallen, also. But God's plan of redemption is going to change all that. Not only is Adam's race going to be redeemed, but all of God's creation is included. God is going to restore everything to its Edenic perfection and original beauty. To say otherwise would be saying that Satan has accomplished his purpose. But that is not true. Jesus came to seek and to save that which was lost and His work is completed. That which was lost in the fall of man will be restored every whit and then some.

Jesus spoke of the "regeneration" in this manner. "Verily I say into you, That ye which have followed Me, in the regeneration when the Son of man shall sit in the throne of His glory, ye shall sit upon twelve thrones, judging the twelve tribes of Israel." (Matt. 19:28.) The Greek word used here is "palengensia," meaning "re-creation" or "making new." It is used only one other time in the Bible, and that is in Titus 3:5, where it speaks of the new birth of the believing person.

We can readily see that the Consummation is not

the end of all things, but, rather, the ending of the World System as we know it. We live in a sin-cursed World System and its sinful character has influenced the whole creation. Paul speaks of this analogy this way, "For I reckon that the sufferings of this present time are not worthy to be compared with the glory which shall be revealed in us. For the earnest expectation of the creature waiteth for the manifestation of the sons of God. For the creature was made subject to vanity, not willingly, but by reason of him who hath subjected the same in hope. For the creature itself also shall be delivered from the bondage of corruption unto the glorious liberty of the children of God. For we know that the whole creation groaneth and travaileth in pain together until now. And not only they, but ourselves also, which have the firstfruits of the Spirit, even we ourselves groan within ourselves, waiting for the adoption, to wit, the redemption of our body." (Rom. 8:18-23.)

Jesus gave many hints of the glory world in His teachings, Once He asked, "Which of you with taking thought can add to his stature one cubit, (about eighteen inches)? If ye then be not able to do that which is least, why take ye thought for the rest?" (Luke 12:26.) Again, while answering the Saducees concerning the resurrection, He said, "They which shall be accounted worthy to obtain that world, and the resurrection from the dead, neither marry nor are given in marriage: Neither can they die any more: for they are equal unto the angels; and are the children of God, being the children of the resurrection." (Luke 20:35,36.) Here, Jesus is saying that we shall be equal to the angels. Matthew and Mark use the term "as the angels," which mean the same.

There is another illustration we can use to expand this wonderful picture and understand more perfectly. The incident is recorded in John, the 20th Chapter. When Mary Magdalene came to the tomb, very early in the morning, she saw the stone rolled away and went and found Peter and John and told them. They immediately went to the tomb themselves to see for themselves. John is very specific in his description of what happened. He said the other disciple, which was he himself, did outrun Peter, but stopped at the entrance of the tomb. Peter ran up behind him and did not stop at the entrance of the tomb, but ran on inside. Then John also went in and beheld how the grave clothes were laying undisturbed. Even the napkin that had been about Jesus' head was still laying in its place. They left the tomb, confused and wondering what had happened. Mary was still standing outside the tomb weeping. When she stooped down and looked inside, what did she see? She saw two men in white standing, one at the head and one at the foot of where Jesus had lain. The question here is, where were these two men when Peter and John entered the tomb? Why did they not see them? The answer is simple! Angels have the power and authority to make themselves visible or invisible. If we shall be equal with the angels, then we, too, shall possess that same power! Not only that, but we shall have a body like unto His glorious body. I John Chapter 3, verses 1 and 2 says, "Behold, what manner of love the Father hath bestowed upon us, that we should be called the sons of God: therefore the world knoweth us not, because it knew Him not. Beloved, now are we the sons of God, and it doth not yet appear what we shall be: but we know that, when He shall appear, we shall be like Him; for we shall see Him as He is." This means what it says! In the resurrection, we shall be not only "equal with the angels," but have a body like His. What a different world the coming Kingdom is going to be! Finite human expressions and descriptions are at a loss and are most inadequate.

According to our text, after the devil was cast into the lake of fire and the wicked dead were resurrected and judged, John saw "a new heaven and a new earth, for the first heaven and the first earth were passed away." What does this mean? The heaven mentioned here is not referring to the dwelling place of God. That will never pass away. That is the third heaven that Paul was caught up to mentioned in II Corinthians, the 12th Chapter. That third heaven will never change. It is perfect as it is. The heaven that John saw as "passing away" was the second heaven. This heaven is where the sun and moon and stars are and the different galaxies. It is the Universe. God said, "Behold, I make all things new." (Rev. 21:5.)

We are now entering into the eternal nature and character of Heaven and Earth and the Universe and all Creation as it shall be. In order to get the fullest picture, we must consider the past along with future eternal aspects. We know that, "By Him (Jesus) were all things created, that are in heaven, and that are in earth, visible and invisible, whether they be thrones, or dominions, or principalities, or powers: all things were created by Him, and for Him: And He is before all thimgs, and by Him all things consist." (Col. 1:16, 17.) This includes the literal Universe. In short, He made everything, both in heaven and on earth, for His glory. He made man, His chief creation, to walk with Him on earth and eventually be with Him in Heaven, the abiding place of God in all His glory. When Adam and Eve sinned in the Garden and fell, all of this was interrupted. But God's sovereign objectives shall be met, every one of them. Complete redemption of all creation shall take place.

I want to refer to the Universe for a moment, as it has been in the past and is now. Let's look at Hebrews, the 1st Chapter, Verses 1 and 2. "God, Who at sundry times and in divers manners spake in times past to the fathers by the prophets, Hath in these last days spoken unto us by His Son, Whom He hath appointed heir of all things, by Whom He made the worlds." Kenneth C. Wuest, the Greek scholar, in his exegesis of these verses gives us some intriguing information. Concerning the word "worlds" in verse two, he makes the following statements. "The word "worlds" is the translation of aionas. The word here includes according to Alford 'God's revelation of Himself in a sphere whose conditions are Time and Space, and so all things existing under these conditions, plus these conditions themselves which exist not independently of the Creator, but are His work, His appointed conditions for all created existence, so that the universe, as well in its great primeval conditions, — the reaches of Space, and the ages of Time, as in all material objects and all successive events, which furnish and people Space and Time, God made by Christ.' The idea in the word aionas is not merely that of the vastness and the magnificence of the physical universe, but the thought of the time and ages through which the purpose and plan of God are gradually unfolding. Thus, the Son is the Divine Agent not only in the original creation of the physical universe, but also in the operation and management of that universe and all its creatures all down the ages of time."

It is this physical Universe as well as the Earth that is going to be renewed or regenerated or re-created at the time of the Consummation. How great is it? We can only guess and even then be far short of an adequate description.

I read recently that scientists had discovered another planet that is five hundred million light years out in space. In trying to understand this distance, we must have some idea of the definition of a light year. It is the distance light can travel in 365 days at the speed of 186,000 miles per second. The distance is mind-boggling. Once you get some idea of how far light can travel in a year, multiply that figure by 500,000,000. Our Lord made all the Universe (and universes if there are others) and it consists by Him and for Him, and for His pleasure they were made. Count the galaxies, if you can, and all the multiplied millions of their suns and moons and stars. He made those, too. In the Re-creation of all things there will be a new heaven (universe) and a new Earth. He will make all things new.

Just suppose we could take a journey through Space and see the expanse of the Universe, and all its sublime beauty and perfection. We would be seeing the physical substance which He has used in His creation or re-creation. These are things that we would be able to touch and handle because they are substantial matter. But there is a sphere or realm that we haven't even considered yet, and that is the spiritual sphere or realm. We will get to that a little later. But the beauty of Eden shall cover the earth and the Universe itself will reflect the wondrous glory of the sovereign, eternal God.

Now, if Eden is to be extended all over the face of the earth, Adam's race will also occupy the earth. That means flesh and blood human beings will still live here, and they shall reproduce and fill the earth with generations of their off-spring. How shall this happen? Let's go back to Armageddon. There was a remnant of God's people that never worshipped the devil nor the antichrist. Although the bulk of the earth's population was wiped out by the commotions that followed Armageddon, this remnant remained true to their God. Consequently, they filled the earth with their generations during the Thousand Year Millennial Reign. When Satan was released from the bottomless pit, he gathered an innumerable army and the Battle of Gog and Magog took place. Everyone did not join with Satan's forces at Gog and Magog. There was a remnant that chose to remain true to Christ, just like the times of Armageddon. This last remnant will constitute the people who will occupy and repopulate the Edenic earth. How do we know this?

It was after John saw the New Jerusalem descending out of heaven from God that this next statement is made. In fact, it is concerning the New Jerusalem. "And the gates of it shall not at all be shut by day: for there shall be no night there. And they shall bring the glory and honor of the nations into it." (Rev. 21: 25,26.) So, you see, nations will still exist. After that there is another statement made akin to this one. "And he showed me a pure river of water of life, clear as crystal, proceeding out of the throne of God and of the Lamb. In the midst of the street of it, and on either side of the river, was there the tree of life, which bare twelve manner of fruits, and yielded her fruit every month: and the leaves of the tree were for the healing of the nations." (Rev. 22:1,2.) Here, the twelve months of the year is mentioned, and the nations still exist. How wonderful is this insight that the Holy Spirit has given concerning the things that await the people of God! The Bible says, "The earth is the Lord's, and the fulness thereof; the world, and they that are therein." (Psalm 24:1.) How wonderful!

When John saw the Holy City, New Jerusalem, coming down from God out of heaven, he gave a vivid description of it. He gave us the measurements of twelve thousand furlongs in every measurement. The length of it was equal to the breadth of it and so was the height. It was the city built foursquare. Actually it was 1500 miles square and 1500 miles high. There is a description of it given in the book, The End Time. I do not know the source of the description. Someone had clipped it out of a paper or magazine and gave it to me while I was writing The End time. I want to share it here.

"John tells us he saw a new heaven and a new earth, for the old heaven and the old earth were passed away. So we know that heaven is going to consist of an entirely new universe in which the redeemed will live. The Bible's description is of a city that "lieth foursquare," with a stated width of 12,000 furlongs. One mathematician did a few calculations. A furlong is equal to an eighth of a mile. So 12,000 furlongs would be about 1,500 miles, the straight line distance both from Florida to Maine, and from New York to the Mississippi River. But that's not all: this four-square city is not only 1,500 miles wide, it is also 1,500 miles HIGH! Why, the Empire

State Building isn't 1,500 FEET high. It's been pointed out that if houses and apartments were made with 15 foot high ceilings, instead of the usual 10 footers, the New Jerusalem would have 528,000 stories, each story would have 2,250,000 square miles. (NOT SQUARE FEET!), and the entire city would have a total area of 1,188,000,000,000 (1 TRILLION, 188 BILLION) square miles of living space. The Department of Eugenics at Carnegie Institute estimates that about 30,000,000,000 (30 BILLION) people have lived on Planet Earth. Now if all of them were to go to heaven, each individual would have a living space of 40 square miles, plenty of room to romp in, I would think. Unfortunately everybody isn't going to heaven. Appropos to this Jesus said, "Enter ye in at the straight gate; for wide is the gate and broad is the way that leadeth to destruction and many there be which go in thereat; Because straight is the gate, and narrow is the way, which leadeth unto life, and few there be that find it." If as many as half of them got there (which I seriously doubt), they'd inherit a piece of property 80 square miles in area. It would be a very long time, indeed, before anybody started hollering for elbow room."

The foregoing could be fairly accurate based upon the measurements of men. I have read more recently that scientists now think there may be as many as 50 billion galaxies in the Universe. Of course, this is just a scientific guess or speculation. There may be even more, but regardless of the number of galaxies, they would all be of physical substance, not spiritual. While it is true that the Holy Spirit moves through space as He did upon the waters in Genesis, the first Chapter, the spiritual realm is quite different. It is intangible and invisible to the physical eye. It will become visible and tangible to us in the new life. We shall see it all in our new bodies after the Rapture. While Job was in his deepest affliction he made a statement as he was moved by the Holy Spirit. He said, "And though after my skin worms destroy this body, yet in my flesh shall I see God." (Job 19:26.) Jesus said, "Blessed are the pure in heart: for they shall see God." (Matt. 5:8.) The Holy Spirit said through Paul, "For now we see through a glass, darkly; but then face to face." (I Cor. 13:12.) Here we are finite creatures, but there, it will be a different world altogether. It will far surpass our most expansive speculations.

Perhaps we can get a fresh glimpse of God's greatness if we consider some of His Divine Attributes. Let's consider His love first of all. I asked this question before, but let me ask it again. When did God first begin to love me, or you? The answer is: He never begun to love us, because He always has loved us. God's love is as eternal as He, Himself. After all, He told Moses at the burning bush, "I shall always be what I always have been." That has never changed and never will. He is the immutable, unchanging God. Let's look at His mercy. Psalm 136 tells us 26 times that His mercy endures forever. Forever is everlasting. There was never a moment in eternity or time that God was unmerciful. His lovingkindness and holiness, His longsuffering and foreknowledge is as eternal as God Himself is.

Let's consider God's foreknowledge. Again, I'll remind you that His foreknowledge has no beginning and it has no end. God does not have to "figure things out" or make guesses about anything. He just knows! He knew all things at the beginning and He knows all things now and He knows all future things. True prophecy is based upon His foreknowledge. His foreknowledge is as perfect as He is. There are no mistakes. That's the reason the Bible firmly declares that "All Scripture is inspired, or God-breathed." Holy men of old spoke , or wrote, as they were moved (carried) by the Holy Ghost. All true prophecy will come to pass because God utters it, based on His perfect foreknowledge. If it is not of God, then it's false prophecy. The unbelief of finite man will not deter nor hinder true prophecy from being fulfilled. Man has nothing to do with it. It is altogether of God! Man's thoughts, negative or positive, have nothing to do with true prophecy. It is from God!

In our consideration of God's divine foreknowledge, we need to include some things about the teachings of predestination. Romans 8:29 says. "For whom He did foreknow, He also did predestinate to be conformed to the image of His Son, that He might be the firstborn among many brethren." In Acts 4:28, the same word translated here "predestinate" is translated "determined." It reads, "For to do whatsoever Thy hand and Thy counsel determined before to be done." This carries the idea of things being "predetermined." In Peter's first pentecostal sermon the Holy Spirit used this same word. Let's hear what He said, for Peter was anointed by the Holy Ghost and literally stood in the presence of God. "Ye men of Israel, hear these words; Jesus of Nazareth, a man approved of God among you by miracles and wonders and signs, which God did by Him in the midst of you, as ye yourselves also know: Him, being delivered by the determinate counsel and foreknowledge of God, ye have taken, and by wicked hands have crucified and slain." (Acts 2:22, 23.) The Greek word here translated "determinate counsel" is the same word that is translated "predestinate" in Romans 8:29 and Ephesians 1:11.

The proper way to interpret Scripture is on a contextual basis. What does the Bible in its context say about a subject? God has said through His Word, "For God so loved the world, that He gave His only begotten Son, that whosoever believeth in Him should not perish, but have everlasting life." (John 3:16.) Peter said in II : Peter 3:9, that "God is not willing that any should perish, but that all should come to repentance." To properly understand what God is saying, we must understand that God has given man a Free Moral Agency, that is, the right to choose—to make his own choice. Since God's foreknowledge is perfect and infinite, He knows the choices each individual will make. For instance, He knew that I would write these lines and thoughts that you are reading right now. And He has always known that you would be reading them right now. God sends no one to Hell! It is the choices a person makes, right or wrong, that determine his or her final and eternal destiny. Our Sovereign God has predetermined before the foundations of the Universe that all who accept His Son as their Saviour will dwell with Him in heaven, and all who do not accept Him will be eternally separated from Him.

How great and wonderful our God is! It is beyond human ability to aptly describe Him or His Divine Attributes. We know that He loves us with an Everlasting Love! The purpose for this short book is to try to understand something of what God has prepared, and is preparing for us. I want to point your attention to two passages of Scripture at this time. In Paul's writings to the Church at Corinth, as the Holy Spirit moved upon him, he wrote, "But as it is written, Eye hath not seen, nor ear heard, neither have entered into the heart of man, the things that God hath prepared for them that love Him. But God hath revealed them unto us by His Spirit: for the Spirit searcheth all things, yea, the deep things of God." (I Cor. 2:9, 10.) Where is this written and who is Paul quoting? Let's look at Isaiah Chapter 64, Verse 4. "For since the beginning of the world men have not heard, nor perceived by the ear, neither hath the eye seen, O God, beside Thee, that He hath prepared for them that waiteth for Him." Let's look at Isaiah, Chapter 65, Verse 17, also. "For, behold, I create new heavens, and a new earth: and the former shall not be remembered nor come into mind." Now, on the basis of these passages and other passages like them, let me wax bold and make some projections.

We know that the last earthly war will take place at the time of Gog and Magog, and that Satan will amass his innumerable army from the population of the Nations which are made up from the generations of the remnant left after Armageddon. There will be a remnant left also after the Battle of Gog and Magog. These will be people who have forever decided to follow the Lamb "whithersoever He goeth." They will be flesh, bone, and blood people. They will be as Adam was before he fell. Actually, they will constitute Adam's race carrying out the command of God to Adam, "Be fruitful, and multiply and replenish the earth (the New Earth) with your "perpetual generations." They will live on the re-created earth and carry out their achievements in the indescribable, brilliant intellect of the first Adam before he fell. They will build their houses, plant their vineyards, and live in the peace and fellowship of God Himself. Earthly seasons will go on and on with twelve months to the year. The four seasons, spring, summer, fall, and winter will still be in play. How else can we understand the Tree Of Life? "The Tree Of Life bares twelve manner of fruits, and yielded her fruit every month: and the leaves of the tree were for the healing of the nations. And there shall be no more curse: but the throne of God and of the Lamb shall be in it; and His servants shall serve Him." (Rev. 22:2, 3.) This describes life as it will be on earth after the re-creation. It will be a perfect Creation as it was before Adam's fall. Isaiah says in Chapter 65, Verse 17 that "the former shall not be remembered, nor come to mind." I take this to mean that sin and its degradation will not have a place in our thoughts. All the disappointments and failures will be forgotten in the light and glorious wonders of the coming world that God has prepared for "them that love Him." We do know this: That His holiness, and His glory will cover the earth like the waters cover the sea.

The Bible says, "I heard a great voice out of heaven saying, Behold, the tabernacle of God is with men, and He will dwell with them, and they shall be His people, and God Himself shall be with them, and be their God." (Rev. 21:3.) In Verse 7 God says, "He that overcometh shall inherit all things, and I will be his God, and he shall be My son." Let me paraphrase these two verses of Scripture. "God will pitch His tent among men, and dwell with them and be their God. They shall look on

Him and say, Father, and He shall look on them and say, Son."What a beautiful thought!

Some think the New Jerusalem will be in the atmosphere above the earth and will be visible to all men on the earth. We do know it will be the dwelling place of the Bride, the Raptured Church. We've been talking mostly about the re-created earth, but what about this New Jerusalem? And what about heaven? How do they fit into the "all things new" picture?

Let us first consider two other questions that come to mind. How long will it take before the earth would become over-populated? And since there will be no death, how will the population be kept in balance? We know according to Jude, the fourteenth verse that Enoch was the seventh generation from Adam. We also know that seven is the number of completeness. Genesis 5:24 says, "And Enoch walked with God; and was not: for God took him." Hebrews, the 11th Chapter, Verse 5 says, "By faith Enoch was translated that he should not see death." Could it be that God is telling us that translation will be the way He will balance the population on the New Earth? Perhaps! It very well could be that post consummation man will walk with God throughout his life and then be translated to heaven as both Enoch and Elijah were. Then, they will have access to heaven and earth as the angels now have.

There are some things on which we can only speculate, but there are others that we can know if we rightly divide the Word of God. We know that in the Incarnation of Jesus Christ there were two natures joined together never to be divided, the very nature of God and the perfect human nature. We know that by virtue of the new birth, we are the offspring of God. "In Him we live, and move, and have our being." (Acts 17:28.) We are heirs of God, and joint heirs with Jesus Christ. Jesus sits at the right hand of God as our Great High Priest. But He is more than that! He is King of Kings and Lord of Lords, not only in heaven but in the earth, also. After His resurrection, He said, "All power is given unto Me in heaven and in earth." (Matt. 28:18.) The inspired Apostle once exclaimed, "All things are yours.....whether the world, or life, or death, or things present, or things to come, all are yours; and ye are Christ's; and Christ is God's." (I Cor. 3:21-23.)

May the grace and peace of God, and our Lord Jesus Christ, and the blessed Holy Spirit be with you all, both now and forever. Amen.